SELF-DRIVING CARS: LEVELS OF AUTOMATION

HEARING

BEFORE THE

SUBCOMMITTEE ON DIGITAL COMMERCE AND
CONSUMER PROTECTION

OF THE

COMMITTEE ON ENERGY AND COMMERCE
HOUSE OF REPRESENTATIVES

ONE HUNDRED FIFTEENTH CONGRESS

FIRST SESSION

MARCH 28, 2017

Serial No. 115–19

Printed for the use of the Committee on Energy and Commerce

energycommerce.house.gov

U.S. GOVERNMENT PUBLISHING OFFICE

25–681 PDF WASHINGTON : 2017

For sale by the Superintendent of Documents, U.S. Government Publishing Office
Internet: bookstore.gpo.gov Phone: toll free (866) 512–1800; DC area (202) 512–1800
Fax: (202) 512–2104 Mail: Stop IDCC, Washington, DC 20402–0001

COMMITTEE ON ENERGY AND COMMERCE

GREG WALDEN, Oregon
Chairman

JOE BARTON, Texas	FRANK PALLONE, JR., New Jersey
Vice Chairman	*Ranking Member*
FRED UPTON, Michigan	BOBBY L. RUSH, Illinois
JOHN SHIMKUS, Illinois	ANNA G. ESHOO, California
TIM MURPHY, Pennsylvania	ELIOT L. ENGEL, New York
MICHAEL C. BURGESS, Texas	GENE GREEN, Texas
MARSHA BLACKBURN, Tennessee	DIANA DeGETTE, Colorado
STEVE SCALISE, Louisiana	MICHAEL F. DOYLE, Pennsylvania
ROBERT E. LATTA, Ohio	JANICE D. SCHAKOWSKY, Illinois
CATHY McMORRIS RODGERS, Washington	G.K. BUTTERFIELD, North Carolina
GREGG HARPER, Mississippi	DORIS O. MATSUI, California
LEONARD LANCE, New Jersey	KATHY CASTOR, Florida
BRETT GUTHRIE, Kentucky	JOHN P. SARBANES, Maryland
PETE OLSON, Texas	JERRY McNERNEY, California
DAVID B. McKINLEY, West Virginia	PETER WELCH, Vermont
ADAM KINZINGER, Illinois	BEN RAY LUJAN, New Mexico
H. MORGAN GRIFFITH, Virginia	PAUL TONKO, New York
GUS M. BILIRAKIS, Florida	YVETTE D. CLARKE, New York
BILL JOHNSON, Ohio	DAVID LOEBSACK, Iowa KURT
BILLY LONG, Missouri	SCHRADER, Oregon
LARRY BUCSHON, Indiana	JOSEPH P. KENNEDY, III, Massachusetts
BILL FLORES, Texas SUSAN	TONY CARDENAS, California
W. BROOKS, Indiana	RAUL RUIZ, California
MARKWAYNE MULLIN, Oklahoma	SCOTT H. PETERS, California
RICHARD HUDSON, North Carolina	DEBBIE DINGELL, Michigan
CHRIS COLLINS, New York	
KEVIN CRAMER, North Dakota	
TIM WALBERG, Michigan	
MIMI WALTERS, California	
RYAN A. COSTELLO, Pennsylvania	
EARL L. "BUDDY" CARTER, Georgia	

SUBCOMMITTEE ON DIGITAL COMMERCE AND CONSUMER PROTECTION

ROBERT E. LATTA, Ohio
Chairman

GREGG HARPER, Mississippi	JANICE D. SCHAKOWSKY, Illinois
Vice Chairman	*Ranking Member*
FRED UPTON, Michigan	BEN RAY LUJAN, New Mexico
MICHAEL C. BURGESS, Texas	YVETTE D. CLARKE, New York
LEONARD LANCE, New Jersey	TONY CARDENAS, California
BRETT GUTHRIE, Kentucky DAVID	DEBBIE DINGELL, Michigan
B. McKINLEY, West Virgina ADAM	DORIS O. MATSUI, California
KINZINGER, Illinois	PETER WELCH, Vermont
GUS M. BILIRAKIS, Florida	JOSEPH P. KENNEDY, III, Massachusetts
LARRY BUCSHON, Indiana	GENE GREEN, Texas
MARKWAYNE MULLIN, Oklahoma	FRANK PALLONE, JR., New Jersey *(ex officio)*
MIMI WALTERS, California	
RYAN A. COSTELLO, Pennsylvania	
GREG WALDEN, Oregon *(ex officio)*	

CONTENTS

	Page
Hon. Robert E. Latta, a Representative in Congress from the State of Ohio, opening statement	1
Prepared statement	3
Hon. Janice D. Schakowsky, a Representative in Congress from the State of Illinois, opening statement	4
Hon. Greg Walden, a Representative in Congress from the State of Oregon, opening statement	5
Prepared statement	6
Hon. Frank Pallone, Jr., a Representative in Congress from the State of New Jersey, opening statement	7
Prepared statement	8

WITNESSES

Jeff Klei, President, North America Automotive Divisions, Continental AG	9
Prepared statement	12
S. William Gouse, Diretor, Federal Programs Development, SAE International	15
Prepared statement	18
David S. Zuby, Executive Vice President and Chief Research Officer, Insurance Institute for Highway Safety	24
Prepared statement	25
Kay Stepper, Ph.D., Vice President for Automated Driving and Driver Assistance Systems, Robert Bosch LLC	34
Prepared statement	36

SUBMITTED MATERIAL

Report, "A Roadmap to Safer Driving Through Advanced Driver Assistance Systems," Motor & Equipment Manufacturers Association,[1] submitted by Mr. Costello

Letter of March 27, 2017, from Jacqueline S. Gillan, President, and Catherine Chase, Vice President of Governmental Affairs, Advocates for Highway & Auto Safety, to Mr. Latta and Ms. Schakowsky, submitted by Mr. Costello 70

Response to Request for Comment, DOT Docket No. NHTSA 092016 090090, Advocates for Highway & Auto Safety, December 2, 2016,[2] submitted by Mr. Costello

Statement of National Safety Council, March 28, 2017, submitted by Mr. Costello 74

Statement of John Bozzella, President and Chief Executive Officer, Global Automakers, March 28, 2017, submitted by Mr. Costello 83

Response to Request for Comment, DOT Docket No. NHTSA 092016 090090, U.S. Chamber of Commerce, November 22, 2016, submitted by Mr. Costello | 87

---

[1] The information has been retained in committee files and also is available at *http://docs.house.gov/meetings/IF/IF17/20170328/105790/HHRG-115-IF17-20170328-SD011.pdf.*

[2] The information has been retained in committee files and also is available at *http://docs.house.gov/meetings/IF/IF17/20170328/105790/HHRG-115-IF17-20170328-SD004.pdf.*

	Page
Statement of American Car Rental Association, March 28, 2017, submitted by Mr. Costello	97
Statement of Dan Galves, Senior Vice President, Chief Communications Officer, Mobileye, March 28, 2017, submitted by Mr. Costello	100
Letter of March 27, 2017, from Marc Rotenberg, President, et al., Electronic Privacy Information Center, to Mr. Latta and Ms. Schakowsky, submitted by Mr. Costello	105
Letter of March 28, 2017, from Honda North America, Inc., to Mr. Latta and Ms. Schakowsky, submitted by Mr. Costello	108

SELF-DRIVING CARS: LEVELS OF AUTOMATION

TUESDAY, MARCH 28, 2017

House of Representatives,
Subcommittee on Digital Commerce and Consumer Protection,
Committee on Energy and Commerce,
Washington, DC.

The subcommittee met, pursuant to call, at 10:05 a.m., in Room 2322 Rayburn House Office Building, Hon. Robert E. Latta (chairman of the subcommittee) presiding.

Members present: Representatives Latta, Harper, Lance, McKinley, Kinzinger, Bilirakis, Mullin, Walters, Costello, Walden (ex officio), Schakowsky, Clarke, Cárdenas, Dingell, Matsui, Welch, Kennedy, Green, and Pallone (ex officio).

Staff present: Ray Baum, Staff Director; Blair Ellis, Press Secretary/Digital Coordinator; Melissa Froelich, Counsel, Digital Commerce and Consumer Protection; Adam Fromm, Director of Outreach and Coalitions; Giulia Giannangeli, Legislative Clerk, Digital Commerce and Consumer Protection/Communications and Technology; Paul Nagle, Chief Counsel, Digital Commerce and Consumer Protection; Olivia Trusty, Professional Staff Member, Digital Commerce and Consumer Protection; Madeline Vey, Policy Coordinator, Digital Commerce and Consumer Protection; Hamlin Wade, Special Advisor for External Affairs; Michelle Ash, Minority Chief Counsel, Digital Commerce and Consumer Protection; Jeff Carroll, Minority Staff Director; Lisa Goldman, Minority Counsel; Caroline Paris-Behr, Minority Policy Analyst; Matt Schumacher, Minority Press Assistant; Andrew Souvall, Minority Director of Communications, Member Services, and Outreach.

Mr. LATTA. Well, good morning. I would like to welcome you all to our subcommittee meeting of the Digital Commerce and Consumer Protection this morning. I really appreciate our witnesses being here. We are going to have members coming in. There is a meeting going on downstairs, and so more folks will be coming in. We see a couple more coming in right now. But I really again appreciate you so for being here, and to get started I would like to recognize myself for 5 minutes for an opening statement.

OPENING STATEMENT OF HON. ROBERT E. LATTA, A REPRESENTATIVE IN CONGRESS FROM THE STATE OF OHIO

Again, good morning. And last month, this subcommittee examined how automakers and other entities are testing self-driving vehicles and preparing for the development of this lifesaving technology. While projections for the development of self-driving vehicles remains years out, advanced driver assistance systems that offer self semi-autonomous driving capabilities are entering the marketplace today.

Advanced driver assistance systems are crash avoidance technologies that can protect drivers, reduce crashes, and enhance the convenience of driving. Forward collision warning, blind spot detection, and lane departure warnings are examples of advanced driver assistance systems. These systems help drivers make safer decisions on the road by providing real-time information about surrounding roadway activity. The driver can receive this information through audible tones, steering wheel vibrations, or small flashing lights on side mirrors alerting the driver of potential safety hazards on the road.

Increasingly, advanced driver assistance systems now entering the market are capable of taking a more active role in the driving task. Innovative systems such as automatic emergency braking and lane departure prevention can temporarily take control over parts of the vehicle's critical safety functions such as braking or steering. This can occur by the system either applying the brakes without input from the driver or steering the vehicle back into marked lanes following unintended drifting.

Automakers and equipment suppliers have announced additional innovative driver assistance systems that are currently in line for deployment. Traffic jam assist can take control of a vehicle's functions in low speed, stop and go traffic. Autonomous valet parking can park itself and retrieve itself when summoned by the owner. And highway autopilot with lane changing is being developed to change lanes and pass other vehicles without the input of the human driver.

The deployment of the advanced driver assistance systems is demonstrating significant safety benefits across the country. Studies are showing that advanced driver assistance systems and crash avoidance technologies are reducing crashes, roadway injuries, and insurance claims. Advanced driver assistance systems are also an essential part in laying the groundwork for the deployment of fully self-driving vehicles.

Through technological advances by manufacturers and equipment suppliers, basic driver assistance systems are taking on more advanced capabilities that assume greater control of the vehicle's critical safety functions throughout a driving trip. The progression of these technologies is incrementally removing the human driver from the driving task and paving the way to full autonomy.

To provide consistency in the development of driver assistance safety technologies, standards-setting organization SAE International developed a classification system to define six different levels of driving automation. SAE levels of automation establish the general scope of the driver assistance system and the role of the human driver in vehicles taking on increasing autonomous driving capabilities.

The levels span from a vehicle with no automation all the way to a vehicle with full automation or a fully self-driving vehicle. Last September, the National Highway Traffic Safety Administration adopted SAE's levels of automation for its own use in its Federal Automated Vehicles Policy.

As we discuss the levels of vehicle automation today, I look forward to learning more about the capabilities of advanced driver assistance systems currently on the market and how these technologies are increasing vehicle safety and protecting America's motorists. I look forward to examining how these systems are informing the development of fully self-driving vehicles and how the auto industry is working to make these systems available across all models and fleets.

I also look forward to hearing from witnesses about how consumers are adopting these technologies and how they are helping to build consumers' confidence in automated driving systems. And with that I will end my opening statement.

[The prepared statement of Mr. Latta follows:]

PREPARED STATEMENT OF HON. ROBERT E. LATTA

Good morning. Last month, this subcommittee examined how automakers and other entities are testing self-driving vehicles and preparing for the deployment of this life-saving technology. While projections for the deployment of self-driving vehicles remain years out, advanced driver assistance systems that offer semi-autonomous driving capabilities are entering the marketplace today.

Advanced driver assistance systems are crash avoidance technologies that can protect drivers, reduce crashes, and enhance the convenience of driving. "Forward Collision Warning," "Blind Spot Detection," and "Lane Departure Warning" are examples of advanced driver assistance systems. These systems help drivers make safer decisions on the road by providing real-time information about surrounding roadway activity. The driver can receive this information through audible tones, steering wheel vibrations, or small flashing lights on side mirrors, alerting the driver to potential safety hazards on the road.

Increasingly, advanced driver assistance systems now entering the market are capable of taking a more active role in the driving task. Innovative systems such as "Automatic Emergency Braking" and "Lane Departure Prevention" can temporarily take control over parts of the vehicle's critical safety functions such as braking or steering. This can occur by the system either applying the brakes without input from the driver or steering the vehicle back into marked lanes following unintended drifting.

Automakers and equipment suppliers have announced additional innovative driver assistance systems that are currently in line for deployment. "Traffic jam assist" can take control of a vehicle's functions in low-speed, stop and go traffic. "Autonomous valet parking" can park itself and retrieve itself when summoned by the owner. And, "highway autopilot with lane changing" is being developed to change lanes and pass other vehicles without the input of a human driver.

The deployment of advanced driver assistance systems is demonstrating significant safety benefits across the country. Studies are showing that advanced driver assistance systems and crash avoidance technologies are reducing crashes, roadway injuries, and insurance claims.

Advanced driver assistance systems are also an essential part in laying the groundwork for the deployment of fully self-driving vehicles. Through technological advancements by manufacturers and equipment suppliers, basic driver assistance systems are taking on more advanced capabilities that assume greater control of the vehicle's critical safety functions throughout a driving trip. The progression of these technologies is incrementally removing the human driver from the driving task and paving the way to full autonomy.

To provide consistency in the development of driver assistance safety technologies, standards-setting organization, SAE (S–A-E) International, developed a classification system that defines six different levels of driving automation. SAE's levels of automation establish the general scope of the driver assistance system and the role of the human driver in vehicles taking on increasing autonomous driving capabilities. The levels span from a vehicle with no automation all the way to a vehicle with full automation or a fully self-driving vehicle. Last September, the National Highway Traffic Safety Administration adopted SAE's levels of automation for its own use in its Federal Automated Vehicles Policy.

As we discuss the levels of vehicle automation today, I look forward to learning more about the capabilities of advanced driver assistance systems currently on the market and how these technologies are increasing vehicle safety and protecting America's motorists. I look forward to examining how these systems are informing the development of fully self-driving vehicles and how the auto industry is working to make these systems available across all models and fleets. I also look forward to hearing from witnesses about how consumers are adopting these technologies and how they are helping to build consumers' confidence in automated driving systems.

Mr. LATTA. I would like to recognize for 5 minutes the gentlelady from Illinois, the ranking member, for 5 minutes. Good morning.

OPENING STATEMENT OF HON. JANICE D. SCHAKOWSKY, A REPRESENTATIVE IN CONGRESS FROM THE STATE OF ILLINOIS

Ms. SCHAKOWSKY. Good morning and thank you, Mr. Chairman and our witnesses. Today's hearing continues our subcommittee's series on autonomous vehicles. In last month's hearing, several of our witnesses referenced different levels of automation and today we will better define those levels and we will also ask about the effectiveness of existing safety technologies.

Self-driving cars are part of a long-term vision to minimize accidents due to human error. Automated features are becoming increasingly common in our cars, but we still have a long way to go to reach full automation, Level 5, as SAE would call it. Technology must be sufficiently tested and ensure that we don't replace human error with system error. In addition, the Takata and Volkswagen scandals raised serious questions about how much we can trust industry to do the right thing on safety.

Volkswagen ordered its supplier to write software to cheat on emissions testing. With software increasingly integral to our vehicles, proper oversight becomes that much more challenging. Ultimately, the success of autonomous features and self-driving cars relies on consumers trusting the technology. Trust must be earned. Once technologies are put in new vehicles it takes decades for technology to become widespread among all vehicles on the road.

Just look at backup cameras. I worked to require backup cameras after I met and talked to parents who were devastated after their children were injured or killed in backover accidents. We passed that law in 2008. Parents and advocates came to DC regularly during the rulemaking process, and NHTSA finally established the standard in 2014. And backup cameras will now be required in all vehicles starting in model year 2018, 10 years after the bill passed.

It will still be years before the passenger vehicles without backup cameras cycle out of use. A car sold today may be on the road for another 2 decades. That is why it is critical we look not only at safety improvements in the long term, but also at which technologies can be effectively deployed right now to save lives.

A lot of safety technologies are out there. However, some are more effective than others. Automatic braking for instance has proven very effective in reducing accidents. The evidence on lane departures systems is more mixed. Today we will hear from the suppliers that develop safety technologies. We will hear about the testing data that is essential to lawmakers as we consider what should be standard, and we will learn about classifying levels of automation, a useful framework as we think about how we move from today's cars to the self-driving cars of the future.

It is a long road ahead, but as I have seen in my years on the subcommittee we have to push forward at every step in the process to make safety improvements a reality. I thank all of our witnesses for being here today, and I look forward to your testimony. And now I would like to yield the remaining time to Representative Matsui.

Ms. MATSUI. Thank you very much, Ranking Member Schakowsky. Innovation and AV vehicle technology is moving at an ever-accelerating pace. We are seeing major investments from traditional auto manufacturers, suppliers like our witnesses from Bosch and Continental, and new entrants like technology companies and ride-sharing platforms. I believe we will make big leaps forward in this space sooner than any of us would have anticipated.

Different companies are pursuing different levels of automation and we know that they do not need to move sequentially through each level of automation. Some companies are choosing to incorporate certain individual features of automation while others are investing in a more integrated Level 4 automation systems today.

In my district in Sacramento we are looking aggressively to the future to lay the foundation for fully autonomous vehicles to be tested on our roads. We are rapidly moving towards a time when truly driverless cars will be on our roads and will coexist with human drivers and other vehicles with different levels of automation.

I look forward to hearing more from our witnesses today and working with all of you to accelerate the testing and deployment of this exciting technology which holds so much promise for improving safety on our roads. I thank you and I yield back.

Ms. SCHAKOWSKY. And I yield back.

Mr. LATTA. Thank you very much. The gentlelady yields back the balance of her time, and at this time the Chair recognizes the gentleman from Oregon, the chairman of the full committee, for 5 minutes for an opening statement.

OPENING STATEMENT OF HON. GREG WALDEN, A REPRESENTATIVE IN CONGRESS FROM THE STATE OF OREGON

Mr. WALDEN. I thank the chairman and I welcome our witnesses and look forward to your delivery of your testimony which I have read and appreciate.

Following years of declining traffic fatalities, we have seen tragically a sharp rise in vehicle-related deaths over the past 2 years. According to early estimates, more than 40,000 Americans, 40,000 people, lost their lives on our Nation's roads last year. That marks a 6 percent increase from 2015. And in my own State of Oregon, 2016 was the deadliest year on the roads in more than a decade, up 20 percent from the year before.

These are sobering numbers. The development of self-driving cars could be a solution to this uptick in danger facing the driving public, the main question is how do we get there? Last month, this subcommittee examined how automakers and other entities are testing self-driving cars and that we are still years away from getting them into hands of consumers.

But that has not stopped the automotive industry from laying the foundation for a complete vehicle autonomy. Today, many cars on the market, including one that my wife owns, are equipped with active safety features or semi-autonomous driving systems. It is pretty impressive to see them in action. These systems have the potential to keep a vehicle within its designated lane; accelerate to pass another vehicle; change lanes, brake, and park all without the input of a human driver.

These advanced driver assistance systems or crash avoidance technologies represent the building blocks to a fully self-driving car. Gradually allowing the vehicle to perform parts of the driving task absent human control means that vehicles are steadily learning how to operate alone and consumers are progressively becoming more familiar and more comfortable with automated driving systems. The advancement of driver assistance systems over the last decade, it is already demonstrating this progression as this technology is minimizing crashes, reducing injuries, and decreasing insurance claims.

In recognition of the safety benefits provided by these systems, the National Highway Traffic Safety Administration has begun to formally incorporate many of these technologies in its 5-Star safety ratings program. Today's hearing will look more closely at many of the advanced driver assistance systems and crash avoidance technologies that are on the road. Our witnesses will also help us to understand the different levels of driving automation, how these technologies are improving safety, and how the development of driver assistance systems and technologies is paving the way for fully self-driving cars.

We often say the development of self-driving cars is a lifesaving endeavor. Following a devastating year on our Nation's roads this could not be any more true. I look forward to a thoughtful and engaging discussion on the levels of driving automation and how advanced driver assistance systems can lead us to the future of a full vehicle autonomy on our road systems.

So thanks for the work you all are doing, thanks for sharing your comments with us. We want to make sure to advance this innovation and technology and save lives on our roads and in our communities.

[The prepared statement of Mr. Walden follows:]

PREPARED STATEMENT OF HON. GREG WALDEN

Following years of declining traffic fatalities, there has been a sharp rise in vehicle-related deaths over the past 2 years. According to early estimates, over 40,000 people lost their lives on our Nation's roads last year, marking a six percent increase from 2015. In Oregon, 2016 was the deadliest year on the roads in more than a decade, up 20 percent from the year before. These are sobering numbers.

The development of self-driving cars could be a solution to this uptick in danger facing the driving public. The main question is: how do we get there?

Last month, this subcommittee examined how automakers and other entities are testing self-driving cars and preparing this innovative safety technology for commercial deployment. Just about everyone concedes that fully self-driving cars are still years away from getting into the hands of consumers; but, that has not stopped the automotive industry from laying the foundation for complete vehicle autonomy.

Today, many cars on the market are equipped with active safety features or semi-autonomous driving systems. These systems have the potential to keep a vehicle within its designated lane; accelerate to pass another vehicle; change lanes; brake; and park—all without the input of a human driver. These advanced driver -assistance systems or crash-avoidance technologies represent the building blocks to a fully self-driving car.

Gradually allowing the vehicle to perform parts of the driving task absent human control means that vehicles are steadily learning how to operate alone and consumers are progressively becoming more familiar and more comfortable with automated driving systems.

The advancement of driver assistance systems over the last decade is already demonstrating this progression, as this technology is minimizing crashes, reducing injuries, and decreasing insurance claims. In recognition of the safety benefits provided by these systems, the National Highway Traffic Safety Administration has begun work to formally incorporate many of these technologies into its 5–Star Safety Ratings program.

Today's hearing will look more closely at many of the advanced driver assistance systems and crash avoidance technologies on the road. Our witnesses will also help us to understand the different levels of driving automation; how these technologies are improving safety; and how the development of driver assistance systems and technologies is paving the way for fully self-driving cars.

We often say that the development of self-driving cars is a life-saving endeavor. Following a devastating year on our Nation's roads, this could not be truer now. I look forward to a thoughtful and engaging discussion on the levels of driving automation and how advanced driver assistance systems can lead us to a future of full vehicle autonomy.

Mr. WALDEN. With that Mr. Chairman, I don't know if anybody else on our side—I would yield to the gentleman from Mississippi for the remainder of my time.

Mr. HARPER. Thank you, Mr. Chairman. Thank you, Chairman Latta, for calling this hearing today to continue the subcommittee's efforts to explore the world of self-driving cars. As I have mentioned at our previous hearings, this topic is of particular interest to me because of the potential opportunities that self-driving cars would provide to Americans with disabilities, including those with intellectual disabilities.

In the disability community lack of transportation is widely viewed as the top impediment to advancement and success in society. Self-driving cars could offer the disability community a new method of transportation to potentially remove this roadblock and provide them additional independence that would open the doors to access new job markets and opportunities to have an even more active role in our society, which benefits us all.

I am looking forward to learning more about the capabilities of advanced driver assistance systems and crash avoidance technologies that are currently on the market and how these capabilities will advance the future of self-driving cars. And with that I yield back.

Mr. LATTA. Thank you. The gentleman yields back, and the Chair now recognizes for a 5-minute opening statement the gentleman from New Jersey, the ranking member of the full committee.

OPENING STATEMENT OF HON. FRANK PALLONE, JR., A REPRESENTATIVE IN CONGRESS FROM THE STATE OF NEW JERSEY

Mr. PALLONE. Thank you, Chairman Latta. Today's hearing gives us a our first true opportunity to talk about what is happening now in automated technology. While learning about the potential technologies of the future is exciting, understanding that there are products currently available that are saving lives and reducing injuries is paramount.

For the foreseeable future, human drivers are going to be driving vehicles on our roads and so efforts to prevent crashes or protect drivers and passengers in a crash are vital. For example, advances such as the addition of airbags and electronic stability control to our cars have saved thousands of lives. As I mentioned at this subcommittee's November hearing on self-driving cars, we see technologies in today's marketplace such as automatic braking that have enormous benefits.

So today I urge all automakers to expedite the deployment of these braking systems into all new vehicles. According to the Highway Loss Data Institute it takes 25 years for a new feature to be on 95 percent of cars on our roads. Therefore, when we see something that works we need to get it on vehicles quickly and it needs to be made standard on all models and makes, not just the most expensive ones.

Witnesses today will discuss other advances such as in lighting and blind spot detection that have promise, and I hope these technologies can help prevent injuries and fatalities. And as with automatic braking, I encourage rapid deployment of any new features that are proven to be beneficial. I also look forward to hearing about research into pedestrian and bicycle rider safety. As we learned at last week's hearing on smart communities, the number of people living in urban areas is rising and those areas have unique transportation challenges.

I am also interested in hearing what new technologies can reduce injuries to rear seat passengers. While injuries to drivers are still the most common, often our most vulnerable passengers are in the back. Unfortunately, data on back seat passengers is still limited which hampers efforts to determine the effectiveness of features intended to protect them.

Therefore, I encourage NHTSA and all other stakeholders to collect and share all relevant data on road safety. We need to be able to see transit opportunities for safety improvements for people riding in the back seats as well as drivers, front seat passengers, and others on the road. More information will also encourage innovation of new safety technologies.

And finally, I will close by continuing my push for security by design and privacy by design where security and privacy are not afterthoughts but built into the products from day 1. I don't think anybody else wants my time, so I will yield back, Mr. Chairman.

[The prepared statement of Mr. Pallone follows:]

PREPARED STATEMENT OF HON. FRANK PALLONE, JR.

Today's hearing gives us our first true opportunity to talk about what is happening now in automotive technology. While learning about the potential technologies of the future is exciting, understanding that there are products currently available that are saving lives and reducing injuries is paramount.

For the foreseeable future, human drivers are going to be driving vehicles on our roads, and so efforts to prevent crashes or protect drivers and passengers in a crash are vital. For example, advances such as the addition of airbags and electronic stability control to our cars have saved thousands of lives.

As I mentioned at this subcommittee's November hearing on self-driving cars, we see technologies in today's marketplace, such as automatic braking, that have enormous benefits. So today, I urge all automakers to expedite the deployment of these braking systems into all new vehicles.

According to the Highway Loss Data Institute, it takes 25 years for a new feature to be on 95 percent of cars on our roads. Therefore, when we see something that works, we need to get it on vehicles quickly and it needs to be made standard on all makes and models, not just the most expensive ones.

Witnesses today will discuss other advances such as in lighting and blind-spot detection that have promise. I hope these technologies can help prevent injuries and fatalities. And as with automatic braking, I encourage rapid deployment of any new features that are proven to be beneficial.

I also look forward to hearing about research into pedestrian and bicycle rider safety. As we learned at last week's hearing on smart communities, the number of people living in urban areas is rising, and those areas have unique transportation challenges. I am also interested in hearing what new technologies can reduce injuries to rear-seat passengers. While injuries to drivers are still the most common, often our most vulnerable passengers are in the back.

Unfortunately, data on back-seat passengers is still limited, which hampers efforts to determine the effectiveness of features intended to protect them. Therefore, I encourage NHTSA, and all other stakeholders, to collect and share all relevant data on road safety. We need to be able to see trends and opportunities for safety improvements, for people riding in the back seats as well as drivers, front seat passengers, and others on the road. More information will also encourage innovation of new safety technologies.

Finally, I will close by continuing my push for "security by design" and "privacy by design," where security and privacy are not afterthoughts but built into the products from day one.

Mr. LATTA. Thank you. The gentleman yields back, and that will conclude our opening statements from our members. The Chair would like to remind Members that, pursuant to committee rules, all Members' opening statements will be made part of the record.

At this time I also want to again thank our witnesses for being with us today. We really appreciate their taking the time to testify before the subcommittee. Today's witnesses will have the opportunity to give opening statements followed by a round of questions from our members.

Our witness panel for today's hearing will include Mr. Jeff Klei, president of Continental Automotive Systems North America at Continental AG; Mr. Bill Gouse, director of Federal Programs at SAE International; Mr. David Zuby, executive vice president and chief research officer at Insurance Institute for Highway Safety; and Dr. Kay Stepper, vice president for Automated Driving and Driver Assistance Systems at Robert Bosch.

We appreciate you all being here with us today and I would like to just mention that we have another subcommittee so we have members coming and out from both subcommittees today. But we look forward to your opening statements and, Mr. Klei, you are recognized for 5 minutes.

STATEMENTS OF JEFF KLEI, PRESIDENT, NORTH AMERICA AUTOMOTIVE DIVISIONS, CONTINENTAL AG; S. WILLIAM GOUSE, DIRECTOR, FEDERAL PROGRAMS DEVELOPMENT, SAE INTERNATIONAL; DAVID S. ZUBY, EXECUTIVE VICE PRESIDENT AND CHIEF RESEARCH OFFICER, INSURANCE INSTITUTE FOR HIGHWAY SAFETY; AND KAY STEPPER, PH.D., VICE PRESIDENT FOR AUTOMATED DRIVING AND DRIVER ASSISTANCE SYSTEMS, ROBERT BOSCH LLC

STATEMENT OF JEFF KLEI

Mr. KLEI. Thank you very much and good morning, Chairman Latta, Ranking Member Schakowsky, and members of the Subcommittee on Digital Commerce and Consumer Protection. I thank the committee for the opportunity to testify today on behalf of Continental. My name is Jeff Klei and I am the president of Continental Automotive Systems in North America.

Continental is a leading tier 1 supplier to develop safe, sustainable, and affordable mobility technology and solutions for our customers. In 2016, we generated more than $43 billion in sales within our automotive tire and specialty rubber groups. Continental employs more than 20,000 employees in the U.S. in more than 80 facilities located in 26 States and has more than 220,000 employees in 55 countries worldwide.

In 2015, there were more than 35,000 lives lost in the U.S. due to traffic crashes. Projections for 2016 are the dismal increase to more than 40,000 fatalities, a level we haven't seen in a decade. More troubling is that on a global scale, roughly 1.2 million people die in roadway crashes and another 50 million are injured each year. This is unacceptable and changing this is what motivates each and every employee at Continental.

In the last 45 years, the U.S. has experienced a relatively declining trend in traffic fatalities due in large part to vehicle safety technology like seatbelts in the '70s, the introduction of anti-lock brake systems and airbags in the '80s, and finally electronic stability control in the '90s. As the auto industry moves towards more widespread implementation of advanced driver assistance systems, Continental projects these technologies will once again reverse the recent increase in fatalities.

Continental and our dedicated employees are committed to developing safe and dynamic driving technologies that contribute to what we call our Vision Zero, a future with zero traffic fatalities, zero injuries, and ultimately zero accidents. Such a future can only be achieved with the help of innovative active and passive safety, advanced driver assistance systems, and automated driving technologies.

With building block technologies like automatic emergency braking, adaptive cruise control, and rear backup assist that are available in vehicles today, we believe we can continue to pursue our Vision Zero and achieve higher levels of automated driving. When we ultimately achieve fully automated driving we believe that we can reduce the number of fatalities by more than 90 percent, the percentage of accidents caused by human error.

The world and the behavior of drivers within it are ever-changing and the vehicles must adapt to these changing trends. Our children seem to rely more on smart phones to stay connected with one another and living in a world of distractions has been commonplace. Automotive technology must develop accordingly.

That is why Continental has put a great deal of effort into human-machine interface technology. We want the driver to be aware of their surroundings, be aware of what systems in the vehicle are doing, and be aware of when it is safe to relinquish control of the vehicle and when it is necessary to re-engage with the vehicle. In addition, we are heavily focused on securing the systems of the vehicle with cybersecurity enhancements as well as the redundancy of safety systems.

Since 2011, we have continued a pursuit of developing and testing highly automated driving with next generation technologies like automated parking, Cruising Chauffeur, and a complete self-driving vehicle in combination with V2X technology. We were the first supplier in the U.S. to be awarded a testing license in the State of Nevada for automated vehicles and are currently testing our third generation automated vehicle on highways and roads throughout the country and around the world.

But our continued efforts in this direction would benefit greatly from an investment in infrastructure that promotes vehicle to X communication, a dedicated spectrum communication band that can be utilized by current and future safety systems, and harmonization of safety laws that allows for the full real world testing of these technologies. The safe commercial deployment of potential lifesaving technology depends on the ability to extensively test on public roads under all conditions.

Finally, we need an update of Federal motor vehicle safety standards to accommodate automated driving technology in a legal framework that supports a new system of mobility. The world of mobility has the capability of expanding to unimaginable independence and personal freedom while enhancing the safety of future generations. Continental stands at the ready alongside our industry colleagues to work with the committee and Congress in helping construct laws and regulations that foster innovation, enable mobility, and create a safer environment for our public.

Thank you again, Chairman Latta, Ranking Member Schakowsky, members of the Subcommittee on Digital Commerce and Consumer Protection, and staff for the opportunity to testify at today's hearing.

[The prepared statement of Mr. Klei follows:]

Testimony of Jeff Klei
President, North America Automotive Divisions
Continental AG

Energy and Commerce Committee
Digital Commerce and Consumer Protection Subcommittee
"Self Driving Cars: Levels of Automation"
March 28, 2017

Good morning Chairman Latta, Ranking Member Schakowsky, and members of the Subcommittee on Digital Commerce and Consumer Protection. I thank the Subcommittee for the opportunity to testify today on behalf of Continental. My name is Jeff Klei, and I am the President of Continental Automotive Systems in North America.

Continental is a leading Tier 1 supplier that develops intelligent technologies for transporting people and their goods. We provide our automotive customers with sustainable, safe and affordable solutions that enhance automotive safety. In 2016 we generated more than $43 billion in sales within our five divisions, Chassis & Safety, Interior, Powertrain, Tires, and ContiTech. Continental employs more than 20,000 employees in the U.S at more than 80 facilities located in 26 states and has more than 220,000 employees in 55 countries worldwide.

In 2015 there were more than 35,000 lives lost in the U.S. due to traffic crashes. Projections for 2016 are expected to increase to more than 40,000 fatalities, a level we haven't seen in a decade. While this is an alarming number, it is even more startling at a global level—more than 1.2 million people die in roadway crashes and another 50 million are injured. This is unacceptable and reversing this trend is what motivates each and every employee at Continental.

In the last 45 years the U.S. has experienced a relative declining trend in traffic fatalities with respect to an increased number of vehicles on the road and number of miles driven. This is due in large part to improved vehicle safety technologies. In the early 1970s the number of injuries and fatalities were at an all-time high. The introduction of the seat belt helped to reduce the total number of traffic fatalities by 10,000 in a few short years. In 1983, the number of fatalities was the lowest in 20 years due to the introduction of anti-lock braking systems. As numbers began to rise again, the airbag became standard in vehicles reducing injuries and fatalities down to its lowest number in 30 years. The introduction of electronic stability control in the mid-1990s helped to reduce traffic accidents to the lowest number in 50 years. Continental projects new crash-avoidance technologies will once again reverse the recent increase in fatalities as the auto industry moves toward a more widespread implementation of Advanced Driver Assist Systems (ADAS).

Innovation has always been at the heart of the automotive industry. From the original concept of the automobile in the late 1800s, the mass production lines pioneered in Detroit, to today, the automotive industry has always invested in research and development to make their products safer, more reliable and more affordable. Today, we are witnessing the automotive industry evolve from a crashworthiness mindset, where manufacturers try to make the passenger cabin more survivable in the event of an accident towards a crash avoidance mindset—after all, the best way to survive a crash is to avoid one in the first place.

Continental, and our dedicated employees, are committed to developing Safe and Dynamic Driving technologies towards Vision Zero. Vision Zero means a future with zero traffic fatalities, injuries and ultimately zero accidents. Such a future can only be achieved with the help of innovative active and passive safety, driver assistance, and automated driving technologies. As Continental brings these technologies to market, we exhaustively test products, and subsystems, as part of a larger system of advanced driving assistance technologies that will be integrated with a variety of components by original equipment manufacturers.

Our Vision Zero philosophy is embedded in each technology we develop as we continue to enable automated driving. At Continental, we describe our systems approach through three primary actions— sense, plan, and act. Whether the technology simply assists the driver like many systems on the road today, or ultimately takes over the driving task completely, it first must *SENSE* the surrounding environment and gather the necessary data that can be interpreted. Sophisticated sensor systems can help eliminate human error and distractions by providing 360-degree awareness of the road at all times. The data gathered from the sensors is then analyzed to identify obstacles or hazards. Our systems then dynamically develop a *PLAN* to determine how to assist the driver. Once that plan is in place, the systems will *ACT* to execute the plan to safely and comfortably pilot the vehicle and in certain cases avoid a hazard or crash situation. Our Sense, Plan, Act approach is the foundation behind Continental's active safety and Advanced Driver Assistance System technology, and is a key component to advancing automated driving systems. We believe that when fully automated driving is possible, traffic fatalities can be reduced by 90 percent because that is the percentage of accidents that are caused by human error.

Continental has been an active participant globally in policy discussions and initiatives with governments, automotive industry partners, trade associations and other standard setting organizations. The collaborative efforts to help establish consistency within the emerging self-driving market has been crucial to the advancement of automated driving technologies. Continental is currently engaged with the Department of Transportation's Smart Cities Program. Several of our divisions are working together to develop a highly sophisticated intersection in Columbus, Ohio, with vehicle and integrated infrastructure technologies that will help save the lives of vehicle occupants as well as pedestrians while improving transportation efficiency in urban environments. We support the National Highway Traffic Safety Administration's recent adoption of the SAE International definitions of automation, as we believe it is beneficial to helping educate the public in order to distinguish between different automated technologies and garner public acceptance.

Continental is one of the leading suppliers in this market, with a complete portfolio of technologies for all defined levels of automation. Each innovative safety feature undergoes an extensive testing process before becoming available to the market. As a supplier, we currently develop a multitude of innovative technologies that can save lives and enhance the driving experience under the Level 0 to Level 2 definitions of automation. These products are designed based on the needs of our customers to assist the driver in interpreting the surrounding environment and control the vehicle in order to prevent an accident from occurring.

Continental has been integral in the deployment of current crash avoidance technologies such as lane keep assist, rear back up assist, automatic emergency braking, and adaptive cruise control, to name a few. These crash avoidance technologies are the building blocks to higher levels of automated driving and need to be embraced as crash avoidance technologies that save lives. All of these technologies can be found throughout the fleets of most vehicle manufacturers.

As the industry moves forward towards Level 3 automation technology and beyond, Continental is positioned to supply public and personal transportation needs with the safest and most advanced technology available on the market. The world and the behavior of drivers within it are ever changing, and the vehicle must adapt to these changing trends. Our children seem to rely on smartphones more so than vehicles. Living in a world of distractions has become commonplace. Automotive technology must be developed accordingly. That is why Continental has put a great deal of effort into Human Machine Interface technology. We want the driver to be aware of their surroundings, be aware of what the systems in the vehicle are doing, and be aware of when it is safe to relinquish control of the vehicle and when to reengage with the vehicle. In addition to informing the occupants, keeping them safe, and pedestrians safe, we must also secure the systems within the vehicle. As part of system development for Highly Automated Driving, we focus on redundancy of vehicle safety systems. That is why we are developing complimentary systems and technologies that support existing safety systems in the vehicle's architecture.

Since 2011, we have continued a pursuit of testing and developing highly automated driving with next generation technologies like automated parking, cruising chauffeur and a complete self-driving vehicle in combination with V2V/V2X technology. We were the first supplier in the U.S. to be awarded a testing license for automated vehicles in Nevada and are currently testing our third generation automated vehicle on highways and roads throughout the country and around the world. We are currently integrating sophisticated technologies such as high resolution flash lidar, which will expand the vehicle's detection capabilities. This is the same technology that has been deployed on space shuttles at the most advanced technical level, and we are working to utilize its potential for road applications. But, our continued efforts in this direction would benefit greatly from an investment in infrastructure that promotes vehicle to X communication, a dedicated spectrum communication band that can be utilized by current and future safety systems, and harmonization of safety laws that allows for the full real world testing of these technologies.

The challenges in broadly testing this new and innovative safety technology across the country are great. The industry currently faces considerable uncertainty on state and federal requirements that would require clarification from the federal government's exclusive authority to regulate all motor vehicles. The safe commercial deployment of potential life saving technology depends on the ability to extensively test on public roads under all conditions. In order to envision a future of full automation, the government must review federal motor vehicle safety standards that would allow for vehicles that may not be under the full control of a driver at all times. Similar to the need of improved road conditions as automobiles transitioned from rural landscapes to metropolitan areas in the early 1900s, we need a road infrastructure that complements automotive advancements, and a legal framework that supports a new system of mobility.

The automotive world is one of excitement. Software developers are becoming automotive suppliers, automotive companies are becoming software developers, and our vehicles are becoming our smart-device. The world of mobility has the capability of expanding to unimaginable independence and personal freedom without sacrificing the safety of future generations. Continental stands at the ready, alongside our industry colleagues, to work with the Committee and Congress in helping construct laws that foster innovation, enable mobility, and create a safer environment for the public.

Thank you again, Chairman Latta, Ranking Member Schakowsky, members of the Subcommittee on Digital Commerce and Consumer protection and staff, for the opportunity to testify at today's hearing.

Mr. LATTA. Well, thank you very much for your testimony today and the Chair recognizes Mr. Gouse for 5 minutes. Thanks again for being here.

STATEMENT OF S. WILLIAM GOUSE

Mr. GOUSE. Thank you, Chairman Latta, Ranking Member Schakowsky, and distinguished members of the subcommittee. SAE International thanks you for the opportunity to participate in this hearing, Self-Driving Cars: Levels of Automation. SAE International is a global society founded in 1905 with more than 140,000 engineers, scientists, related technical experts, and students in over a hundred countries in the aerospace, automotive, motorcycle, commercial, construction, agricultural, and specialized vehicle industries.

Some notable members were aviation and automotive pioneers Orville Wright, Henry Ford, Amelia Earhart and Ransom Olds, motor sports legends such as Andy Granatelli and Dan Gurney, along with celebrities like Jay Leno. SAE members from Government, academia, and industry have testified at this subcommittee or at previous hearings in both chambers. All four of us on this panel today testifying are SAE members, as I see are many of my colleagues in the audience.

My SAE experience began even before I was a freshman mechanical engineering student at Georgia Tech when my professor and SAE Student Chapter advisor Professor Williams signed me up as a student member and gave me this membership pin. My initial exposure to SAE was before college because my father was or actually still is an SAE member.

SAE's core competencies are voluntary consensus standards development with nearly 30,000 experts across the globe contributing to a continually growing standards portfolio of over 10,000 active and 25,000 historical standards. These are used to increase safety, performance, quality and productivity of personal commercial transportation services while optimizing cost of products and product life cycles. This is an important point as this standard I will discuss in more detail in a moment is a product, as all standards are, of our members and other volunteers' efforts.

In addition to the standards activities SAE holds dozen of conferences and symposia, including the Government/industry meeting held in January in conjunction with the Washington Auto Show Mobility Talks, and next week is the SAE World Congress in Detroit where my colleagues are also presenting and participating. These events plus other mutually beneficial Government/industry academic networking opportunities provide information for the formation of sound public policy positions and affiliated programs, products, and services that add value and encourage innovation.

SAE standards are referenced in Government regulations, procurement documents, recommendations, and guidelines issued by the U.S. DOT, the U.S. EPA, Department of Energy, the NTSB, in regulations in our States, Commonwealths, inhabited territories, and local jurisdictions. In addition, SAE standards are used internationally, Canada, elsewhere in the Americas, overseas, and by the UNECE.

SAE believes that incorporating voluntary consensus standards by reference as directed in the National Technology Transfer Advancement Act and the Office of Management and Budget Circular-A119 improves the efficiency and effectiveness of Government, whether a Federal, State, municipal body, or global harmonization activity, it saves time and money while increasing the efficacy of policy, legislation, and/or regulation. This is critical in order to respond to the policy or regulatory needs brought about by the rapid technology developments we are witnessing.

These developments are progressing significantly faster, potentially orders of magnitude faster, than the regulatory process. In addition, the competitiveness of products and services increased in the global marketplace because of the higher quality, value, and customer confidence achieved through conformity with SAE standards. SAE has several standards published and many documents in development by a variety of car, motorcycle, pedestrian, and truck and bus committees relating to increasing the safety and efficiency of transport.

While work continues to improve passive safety and crashworthiness of vehicles, the potential of implementing technological solutions to avoid or reduce the severity of crashes is a major focus of our SAE committee activities. Details of these efforts, standards and documents, and progress were submitted to the subcommittee in written testimony. In summary, they encompass active safety systems, driver assistance systems, cybersecurity, vehicle connectivity and communications, measurement and test devices, vehicle testing including safe on-road testing of automated driving systems, and specific to today's hearing, title SAE International Standard J3016: Taxonomy and Definitions for Terms Related to Automated Driving. I believe there is a flyer in front of all of you of this standard.

This recommended practice originally published in 2014 and revised last September and referenced in the Federal Automated Vehicles Policy provides stakeholders including Federal, State, and local/municipal regulators, policy makers with a taxonomy describing the full range of six levels, SAE 0 through 5, of driving automation in on-road motor vehicles. These six levels span from no automation to full automation.

I want to point out the key distinction. You see a dark green break in the handout here is between Level 2 where the human driver performs part of the dynamic driving task and Level 3 where the automated driving system performs the entire dynamic driving task under various conditions. The document, J3016, also contains functional definitions for advanced levels of driving automation and over a dozen related terms and definitions.

Additional terms and definitions of active safety systems are contained in another standard, J3063 that was published in November of 2015. Importantly, what these standards do not provide are specifications or otherwise imposed requirements on driving automation systems or active safety systems, nor does it imply any particular order of market introduction or adoption. One vehicle might have multiple driving automation features such that it could operate at or different levels depending upon the features that are engaged or other consideration.

Standardizing levels of driving automation and supporting terms serve several purposes particularly clarifying the role of the human driver, if any, during driving automation system engagement; providing a useful framework for driving automation specifications and technical requirements; providing clarity, consistency, and stability in communications on the topic of driving automation, as well as a useful shorthand that saves considerable effort and time. The document is designed to be useful to many beyond the engineering community, such as legislators, regulators, others in the legal profession, the general and trade media, and consumers and the public that are buying, riding in, or having freight delivered in a vehicle with some level of driver assistance or automation.

The levels I will go through very briefly are 0, with no automation; 1, a driver assistance system to a specific mode such as keeping steering or accelerating/decelerating; Level 2, partial automation, one or more driver assistance systems, both steering and acceleration/deceleration using information about the driving environment. The human driver is still expected to perform all remaining aspects.

That break down to automated driving systems that monitor the driving environment for Level 3 conditional automation, driving mode-specific performed by an automated driving system in all aspects of the dynamic driving task which define the standard, with the expectation that the human driver will respond appropriately with a request to intervene; 4, high automation, the driving mode-specific performance by an automated driving system of all aspects of the driving task even if a human driver does not respond appropriately to a request to intervene; and 5, full automation, full-time performance by an automated driving system of all aspects of the dynamic driving task under all roadway and environmental conditions that can be managed by a human driver.

SAE has been and will continue to work with organizations and entities to reference SAE standards as we learn of their policy, regulatory, and legislative activities regarding both the public on-road testing, and the deployment of vehicles with driver assistance and automation systems. We are members of the Commonwealth of Pennsylvania——

Mr. LATTA. Pardon me, Mr. Gouse, if you could just wrap up, please.

Mr. GOUSE. All right. We are members of the Pennsylvania Department of Transportation Task Force; we work with the associated motor vehicle manufacturers and other groups. SAE levels of automation were adopted in the Declaration of Amsterdam and they are used as we spoke earlier of the U.S. DOT and the Federal Automated Vehicles Policy. Prior to this, the Government used separate terms and retired their classification so now we have this consistent usage.

Driving assistance and automated driving systems have the potential to provide substantial benefits to all customers of road transport. And I thank you very much for this opportunity to provide this statement and answer any questions.

[The prepared statement of Mr. Gouse follows:]

United States House of Representatives

Committee on Energy and Commerce

Subcommittee on Digital Commerce and Consumer Protection

Self-Driving Cars: Levels of Automation Hearing

March 28, 2017

Statement of SAE International

S. William Gouse, Director, Federal Program Development

Chairman Latta, Ranking Member Schakowsky, and distinguished members of the subcommittee, SAE International thanks you for the opportunity to submit this statement for the hearing: "Self-Driving Cars: Levels of Automation".

SAE International is a global society founded in 1905 with more than 140,000 engineers, scientists, related technical experts, and students, in over 100 countries, in the aerospace, automotive, motorcycle, commercial, construction, agricultural and specialized vehicles industries. Some notable members were aviation and automotive pioneers Orville Wright, Henry Ford, Amelia Earhart, and Ransom Olds, and celebrities such as Andy Granatelli and Jay Leno. SAE members from government, academia, and industry have testified to this subcommittee at previous hearings.

SAE's core competencies are voluntary consensus standards development with nearly 30,000 experts across the globe contributing to a continually growing standards portfolio of over 10,000 active and 25,000 historical standards used to increase safety, performance, quality, productivity of personal commercial transportation services while optimizing cost of products and product-life cycles. SAE holds dozens of conferences and symposia, including the Government Industry meeting held in January in conjunction with the Washington Auto Show and our World Congress next month in Detroit. These events, plus other mutually beneficial government/industry academic networking opportunities provide information for the formation of sound public policy positions and affiliated programs, products and services that add value and encourage innovation.

SAE standards are referenced in government procurement documents, recommendations, guidelines, and in regulations issued by the US Department of Transportation (US DOT), US Environmental Protection Agency, US Department of Energy, the National Transportation Safety Board, in regulations in our states, commonwealths, and inhabited territories. In addition, SAE standards are used internationally, in Canada, elsewhere in the Americas, overseas and by the United Nations Economic Commission for Europe. SAE believes that incorporating voluntary consensus standards by reference as directed in the National Technology Transfer Advancement Act and in Office of Management and Budget Circular-A119 improves the efficiency and effectiveness of government (whether a Federal, state, municipal body, or global harmonization activity), saving time and money, while increasing the efficacy of policy, legislation, and/or regulation. This is critical in order to respond to the policy or regulatory needs brought about by rapid technology developments. These developments are progressing significantly faster, potentially orders of magnitude faster, than the regulatory process. In addition, the competitiveness of adopting organizations' products and services is increased in the global marketplace because of the higher quality, value, and customer confidence achieved through conformity with SAE standards.

SAE has several standards published and many documents in development by a variety of car, motorcycle, pedestrian, and truck and bus committees relating to increasing safety and efficiency of transport. While work continues to improve passive safety and crashworthiness of vehicles, the potential of implementing technological solutions to avoid or reduce the severity of crashes is a major focus of our members' committee activities. These efforts encompass active safety systems, driver assistances systems, cybersecurity, vehicle connectivity and communications, measurement and test devices, vehicle testing including safe on-road testing of automated driving systems, and specific to today's hearing's title, SAE International's standard J3016: Taxonomy and Definitions for Terms Related to On-Road Motor Vehicle Automated Driving Systems.

This Recommended Practice originally published in 2014 and revised last September, and referenced in the Federal Automated Vehicles Policy provides stakeholders including Federal, state, and local/municipal legislators, regulators and policy-makers with a taxonomy describing the full range of levels (SAE 0 through 5) of *driving automation* in on-road *motor vehicles*. This includes functional definitions for advanced levels of *driving automation* and related terms and definitions. This Recommended Practice does not provide specifications, or otherwise impose requirements on, *driving automation systems*. Standardizing levels of *driving automation* and supporting terms serves several purposes, including:

• Clarifying the role of the (human) *driver*, if any, during *driving automation system* engagement.

• Answering questions of scope when it comes to developing laws, policies, regulations, and standards.

• Providing a useful framework for *driving automation* specifications and technical requirements.

• Providing clarity and stability in communications on the topic of *driving automation*, as well as a useful short-hand that saves considerable time and effort.

This document was developed per the following guiding principles:

• Be descriptive and informative rather than normative.

- Be consistent with current industry practice.

- Be consistent with prior art to the extent practicable.

- Be useful across disciplines, including engineering, law, media, public discourse.

- Be clear and cogent, provide functional definitions and avoid or define ambiguous terms.

SAE International's Levels Of Automated Driving Automation For On-Road Vehicles:

SAE level	Name	Narrative Definition	Execution of Steering and Acceleration/ Deceleration	Monitoring of Driving Environment	Fallback Performance of Dynamic Driving Task	System Capability (Driving Modes)
Human driver monitors the driving environment						
0	no Automation	the full-time performance by the *human driver* of all aspects of the *dynamic driving task*, even when enhanced by warning or intervention systems	Human driver	Human driver	Human driver	n/a
1	Driver Assistance	the *driving mode*-specific execution by a driver assistance system of either steering or acceleration/deceleration using information about the driving environment and with the expectation that the *human driver* perform all remaining aspects of the *dynamic driving task*	Human driver and system	Human driver	Human driver	Some driving modes
2	Partial Automation	the *driving mode*-specific execution by one or more driver assistance systems of both steering and acceleration/deceleration using information about the driving environment and with the expectation that the *human driver* perform all remaining aspects of the *dynamic driving task*	System	Human driver	Human driver	Some driving modes
Automated driving system ("system") monitors the driving environment						
3	conditional Automation	the *driving mode*-specific performance by an *automated driving system* of all aspects of the dynamic driving task with the expectation that the *human driver* will respond appropriately to a *request to intervene*	System	System	Human driver	Some driving modes
4	high Automation	the *driving mode*-specific performance by an automated driving system of all aspects of the *dynamic driving task*, even if a *human driver* does not respond appropriately to a *request to intervene*	System	System	System	Some driving modes
5	full Automation	the full-time performance by an *automated driving system* of all aspects of the *dynamic driving task* under all roadway and environmental conditions that can be managed by a *human driver*	System	System	System	All driving modes

SAE has been and will continue to work with states and commonwealths as we learn of their policy, regulatory, and legislative activities regarding both the public on-road testing and the deployment of vehicles with driver assistance and automated driving systems. For example, we have been working the American Association of Motor Vehicle Administrators members and staff, and participate and contribute to their "Automated Vehicles Information Sharing Group". We're also members of the Commonwealth of Pennsylvania's Department of Transportation Autonomous Vehicle Testing Policy Task Force and continue to work with several states regarding policy and regulations for testing of automated driving systems on public roads. SAE staff and members have been involved with efforts in California and Michigan, but many more states, commonwealths, and municipalities in the US are developing policy, regulations and legislation in their departments of transportation, motor vehicle administrations, and legislative bodies.

SAE was pleased that the US DOT and the National Highway Traffic and Safety Administration (NHTSA) adopted the SAE levels of automation in their September 2016 "Federal Automated Vehicles Policy". Government entities, media, trade and consumer associations and the transportation/vehicle industry were using SAE levels of automation or the differing NHTSA classification and some, both. This incongruent usage has been one of the difficulties with communicating with stakeholders and harmonizing policy efforts. An ongoing challenge is to expand the use of consistent terms, definitions, and procedures via increasing the awareness of SAE's resources, such as SAE J3016, SAE J3018: Guidelines for Safe On-Road Testing of SAE Level 3, 4, and 5 Prototype Automated Driving Systems (ADS), and other SAE International's enabling standards and recommended practices to all levels and branches of government.

Driver assistance and automated driving systems have the potential to provide substantial benefits to all customers of road transport. SAE International's wealth of knowledge and experience with existing and developing standards in this area should be used by government, industry and academia to assist in the technology and policy solutions necessary for the adoption of these life and property saving advances.

SAE International thanks the Subcommittee for this opportunity to provide a statement and welcomes questions and requests for additional information.

Supplementing this statement, SAE International submits the following documents for the record:

Surface Vehicle Recommended Practice, SAE J3016, Taxonomy and Definitions for Terms Related to Automated Driving Systems, issued January 2014, revised September 2016.

Automated & Connected Transportation Standards, An infographic from SAE International

Automated Driving, Levels of Driving Automation are Defined in New SAE International Standard J3016

SAE International Global Ground Vehicle Standards Organization Chart

Mr. LATTA. Well, thank you very much. And Mr. Zuby you are recognized for 5 minutes and thank you very much for being here.

STATEMENT OF DAVID S. ZUBY

Mr. ZUBY. Good morning, Chairman Latta, Ranking Member Schakowsky, and distinguished members of the subcommittee. On behalf of the Insurance Institute for Highway Safety, thank you for the opportunity to testify today on vehicle automation and crash avoidance technologies.

The Insurance Institute for Highway Safety and its sister organization, the Highway Loss Data Institute, are nonprofit research institutes that identify ways to reduce deaths, injuries, and property damage on our highways. We are wholly supported by voluntary contributions from companies that sell automobile insurance in the United States and Canada.

The United States has made enormous progress in reducing the toll from motor vehicle crashes. The death rate per billion vehicle miles traveled is one quarter of what it was in 1973 when crash deaths peaked at 54,589. While changes in traffic laws and their enforcement combined with changes in road and vehicle designs all contributed to that decline, our research has shown that improvements in vehicle safety have been the largest contributor to road safety since the 1990s. We are convinced that further improvement in vehicle safety will remain an important strategy to make travel on U.S. roads even more safe in the future.

Past improvements in vehicle safety largely focused on mitigating and preventing injuries when crashes occurred. The newest tool in the vehicle safety toolbox is automation of the vehicle controls that can prevent crashes in the first place and reduce the severity of those that aren't prevented. Electronic stability control which helps prevent sideways skidding and loss of control, reduces the risk of a fatal single vehicle crash by 49 percent and cuts the risk of a fatal multiple vehicle crash by 20 percent.

More recently, front crash prevention systems which help drivers avoid front to rear crashes with warnings or automatic braking reduce these crashes by 26 percent for warnings by itself and by 50 percent for warnings combined with auto braking. Reductions for crashes with injuries are even larger.

These are large reductions and count as wins for automation of vehicle control, but neither ESC nor front crash prevention systems prevent all the crashes they target. In addition, there are other new crash avoidance technologies like those that aim to prevent crashes precipitated by inadvertent lane drifts for which we have not yet found definitive benefits. There are reasons to be skeptical of the claims that driving automation will eliminate all crashes currently caused by human error. This is especially true in the near term technologies which will continue to involve human driver to a large extent.

The design of these technologies and how drivers interact with them will be an important factor in their success. For example, we have found that on average across multiple implementations from various automakers, lane departure warning and other lane maintenance systems are used by only 50 percent of drivers whose cars have them. There is a wide variation in the use rate and that seems to be influenced by system design.

As technology allows further automation of the driving task, we are concerned that some human drivers will fail to understand the limitations of these systems on their vehicles and crash because they are overly reliant on them. The design of driving automation systems will be key to helping drivers understand how systems work including the limitations of the technology. It will be important to continually monitor the effects of safety on new technologies entering the market.

The studies mentioned above were only possible with close cooperation of a few automakers who helped us identify by vehicle identification number the specific vehicles that were equipped with a range of optional features. Unfortunately, there was no comprehensive database linking VINs to information about what features are present on a given vehicle. Government policies aimed at ensuring the availability of such highway safety data are important to enhance highway safety research on the effectiveness of these emerging technologies.

Thank you again to the members of the subcommittee for inviting me to share what IIHS and HLDI have learned about the effectiveness of crash avoidance technologies. I would be happy to answer any questions.

[The prepared statement of Mr. Zuby follows:]

Statement before the United States House of Representatives Committee on Energy and Commerce; Subcommittee on Digital Commerce and Consumer Protection

What today's advanced driver assistance systems can tell us about the self-driving future

March 28, 2017

David S. Zuby
Insurance Institute for Highway Safety

1005 N. Glebe Road, Suite 800
Arlington, VA 22201
+1 703 247 1500

iihs.org

Summary

The United States has made enormous progress in reducing the toll from motor vehicle crashes, thanks to safer vehicles, better laws and enforcement, and traffic engineering improvements. Of those factors, vehicle improvements have played the biggest role in recent years. In contrast, efforts to reduce crashes by changing driver behavior have largely stalled.

Automation is the next frontier in vehicle improvements and could one day address the problem of human behavior by taking it out of the equation completely. That day remains far in the future, however.

Experiences with existing crash avoidance technologies can give us some clues regarding the potential benefits and pitfalls of emerging automation technologies. IIHS research has documented safety benefits from some features, including electronic stability control and automatic braking. On the other hand, studies of insurance claims have not found consistent benefits from lane departure warning systems. These results show how crucial it will be to monitor new technologies to see if they deliver on their promise. Policies to help ensure the availability of information about which specific vehicles are equipped with which features would help researchers track the effectiveness of driver assistance systems.

Driver attitudes toward technologies will be key to ensuring new features reach their potential. Our research has shown that driver acceptance of technology varies.

We expect driving automation to enter the market gradually. During these years of technical evolution, some drivers may fail to understand the limitations of the systems and become overly reliant on them. New features should be designed in such a way as to make their limitations clear.

While automation has the potential to greatly reduce the toll from crashes, it would be a mistake to focus on it to the exclusion of proven countermeasures. Things like lower speed limits and strict enforcement of seat belt laws can provide benefits now, while we await the self-driving future.

Introduction

The Insurance Institute for Highway Safety (IIHS) and its sister organization, the Highway Loss Data Institute (HLDI), are nonprofit research institutes that identify ways to reduce deaths, injuries, and property damage on our highways. We are wholly supported by voluntary contributions from companies that sell automobile insurance in the U.S. and Canada. Thank you for the opportunity to testify on emerging automated driving technologies.

The United States has made enormous progress in reducing motor vehicle crash deaths over the past half a century. A combination of safer vehicles, better laws and enforcement of those laws, and traffic engineering improvements have cut the rate of crash deaths per population to nearly half of what it was in 1975.[1] The rate of crash fatalities per 100 million vehicle miles traveled is one-third the rate in 1980.

Out of all these types of countermeasures, it is vehicle improvements — including more crashworthy structures, front and side airbags and electronic stability control (ESC) — that have driven most of the decline in driver death rates since the mid-1990s.[2] In contrast, efforts to reduce crashes by changing driver behavior have largely stalled. Speeding, alcohol-impaired driving and lack of safety belt use all remain persistent problems.

Automation is the next frontier in vehicle improvements and could also address the problem of driver behavior. Full automation has the potential to make the human propensity to make poor decisions and errors irrelevant. In a study of police-reported crashes occurring during 2005-07 where at least one vehicle was towed from the scene, researchers found that a driver's error or physical state had led to 94 percent of the crashes.[3] If automation can eliminate all crashes involving driver-related factors, then thousands of lives will be saved each year.

At the moment and for the foreseeable future, however, human drivers are still a key part of the equation. The safety potential of partial automation will be limited in large part by the way human drivers interact with driver assistance systems on their own vehicles and with fully automated vehicles with which they may share the road.

What we can learn from existing crash avoidance features

Although full driving automation for most vehicles remains far in the future, crash avoidance features that automatically assume control over vehicle motion when drivers fail to adequately respond to crash hazards aren't new. These include ESC and automatic braking systems. Our research has already documented injury-preventing benefits of these features.

ESC, which has been required on all new passenger vehicles since the 2012 model year, helps prevents sideways skidding and loss of control. The technology reduces the risk of a fatal single-vehicle crash by 49 percent and cuts the risk of a fatal multiple-vehicle crash by 20 percent for cars and SUVs.[4] Its effectiveness in preventing rollover crashes is even more dramatic. Years ago, SUVs were considered dangerous vehicles because their high centers of gravity made them prone to rolling over. That is no longer the case, thanks to ESC, which reducing the risk of fatal single-vehicle rollover crashes by 75 percent for SUVs and by 72 percent for cars.[4]

More recently, automatic control of vehicle brakes has proven to be an effective countermeasure against front-to-rear crashes. Front crash prevention is our name for systems that can detect an impending collision with the vehicle in front and warns the driver to brake, automatically brakes on its own or performs a combination of these functions. In a study of police-reported front-to-rear crashes, we found that systems with automatic braking reduce rear-end crashes by about 50 percent.[5] Studies by HLDI of insurance claim rates have also shown benefits for front crash prevention systems with and without automatic braking.[6,7,8,9,10,11]

Despite these success stories, not all crash avoidance features have been shown to be effective. For example, HLDI examined the effectiveness of lane departure warning systems from six manufacturers and did not find any consistent changes in rates of insurance claims covering damage to at-fault vehicles, which is the type of claim that would likely follow a single-vehicle run-off-road crash.[7,9,12,13]

The disparate results for the effects of crash avoidance technologies point to one of our concerns about driving automation — namely, that there is no guarantee that the technology will deliver on its promise. Consequently, it will be important to continually monitor the effects on safety of new technologies entering the market. The studies mentioned above were only possible with the close cooperation of a few automakers who helped us identify by Vehicle Identification Number (VIN) the specific vehicles that were equipped with a range of optional features. Unfortunately, there is no comprehensive database linking VINs to information about what features are present on a given vehicle. Government policies aimed at ensuring the availability of such data for highway safety research would greatly enhance our ability to study the effectiveness of emerging technologies.

Driver attitudes

Collision avoidance and driving automation systems can't reach their crash-reduction potential if drivers don't use or respond appropriately to them. A recent IIHS observational study illustrates how driver attitudes toward advanced driver assistance systems can vary depending on how the feature is implemented.[14] We observed vehicles from eight manufacturers brought to dealership service centers to see if their front crash prevention and lane-maintenance systems (i.e., lane-departure warning, lane-departure prevention or active lane-keeping) were turned on. While front crash prevention was activated in 93 percent of vehicles we observed, lane-maintenance systems were turned on in only 51 percent of vehicles.

We also studied driver trust in advanced technologies in a more direct way by inviting our own employees to drive vehicles equipped with adaptive cruise control, forward collision warning, lane-departure warning, active lane-keeping and side-view assist systems. Fifty-four employees took part in this study, using the vehicles for days or weeks at a time for both commuting and longer trips. Overall, drivers did not express strong trust in any of the technologies.[15] Trust was highest for side-view assist and lowest for active lane-keeping. Trust in adaptive cruise control and side-view assist varied among vehicles.

Pitfalls of partial automation

No matter how quickly technology develops, it will take at least 25 years before nearly all vehicles on U.S. roads have today's latest technology. This estimate is based on a HLDI study that examined how long it takes for new features to be present in 95 percent of registered vehicles.[16] Thus, if the government were to require that all new vehicles sold in the U.S. be fully automated starting tomorrow, it would still be 2042 before nearly all vehicles on the roads were fully automated.

More realistically, we think driving automation will enter the market in a piecemeal fashion. Over time more and more of the driving task will be able to be automated. During these years of technical evolution, we are concerned that some human drivers will fail to understand the limitations of the systems on their vehicles and crash because they are overly reliant on them. Driving automation systems should be designed in ways that make their limitations clear to human operators.

It is also worth noting that partial automation may be of limited benefit in many kinds of crashes. We recently examined records of crashes caused by drivers drifting from their lanes. We found that 34 percent of drivers in lane-drift crashes were asleep or otherwise incapacitated because of a medical issue or alcohol or drug use.[17] For those drivers, lane-maintenance systems would have little relevance. Even if these vehicles had been brought back into their lanes, they likely would have crashed ultimately. To be effective in such cases, a crash avoidance system would have to bring the vehicle to a stop on the side of the road.

Finally, there is the issue of autonomous vehicles sharing the road with human drivers. Our study of crashes on public roads involving Google's self-driving cars shows that even high-performing self-driving vehicles will still be struck by vehicles driven by humans.[18] We reviewed 19 crashes involving Google self-driving cars traveling in autonomous mode. In most of the incidents, the Google car was rear-ended by another vehicle.

Other opportunities to reduce crash deaths and injuries

Our work at IIHS and HLDI is guided by a rubric known as the Haddon matrix. Developed by William Haddon Jr., the nation's first highway safety chief and president of IIHS from 1969 to 1985, the matrix reminds public health practitioners and policymakers that there are often multiple opportunities to treat a public health problem such as motor vehicle crashes.

Improvements in vehicle safety have been effective in reducing crash deaths in recent decades, and increasing automation is the next logical step in those efforts. However, it would be a mistake to focus on those opportunities to the exclusion of proven countermeasures. Lower speed limits, strict enforcement of seat belt laws and prohibitions on alcohol-impaired driving, and safer road designs are just some of the tools that could be used to reduce the toll from crashes while we wait for the benefits of driving automation.

References

1. Insurance Institute for Highway Safety. Yearly snapshot 2015. http://www.iihs.org/iihs/topics/t/general-statistics/fatalityfacts/overview-of-fatality-facts. Accessed on March 23, 2017.
2. Farmer, C. M.; Lund, A.K. 2015. The effects of vehicle redesign on the risk of driver death. *Traffic Injury Prevention* 16(7): 684-690.
3. Singh, S. 2015. Critical reasons for crashes investigated in the National Motor Vehicle Crash Causation Survey. Traffic Safety Facts, Report no. DOT HS 812 115. Washington, DC: National Highway Traffic Safety Administration.
4. Farmer, C. M. 2010. Effects of electronic stability control on fatal crash risk. Arlington, VA: Insurance Institute for Highway Safety.
5. Cicchino, J. B. 2017. Effectiveness of forward collision warning and autonomous emergency braking systems in reducing front-to-rear crash rates. *Accident Analysis and Prevention* 99(Part A): 142-152.
6. Highway Loss Data Institute. 2016. Fiat Chrysler collision avoidance features: initial results. *Loss Bulletin* 33(2). Arlington, VA.
7. Highway Loss Data Institute. 2016. 2013-15 Honda Accord collision avoidance features. *Loss Bulletin* 33(32). Arlington, VA.
8. Highway Loss Data Institute. 2016. Mercedes-Benz collision avoidance features — a 2016 update. *Loss Bulletin* 33(23). Arlington, VA.
9. Highway Loss Data Institute. 2012. Volvo collision avoidance features: initial results. *Loss Bulletin* 29(5). Arlington, VA.
10. Highway Loss Data Institute. 2016. Acura collision avoidance features — a 2016 update. *Loss Bulletin* 33(19). Arlington, VA.
11. Highway Loss Data Institute. 2016. 2013-15 Subaru collision avoidance features. *Loss Bulletin* 33(30). Arlington, VA.
12. Highway Loss Data Institute. 2012. Mercedes-Benz collision avoidance features: initial results. *HLDI Bulletin* 29(7). Arlington, VA.
13. Highway Loss Data Institute. 2016. Mazda collision avoidance features: an update. *HLDI Bulletin* 33(3). Arlington, VA.
14. Reagan, I.J.; Cicchino, J.B.; Kerfoot, L. B.; and Weast, R.A. 2017. Crash avoidance and driver assistance system technologies – are they used? Arlington, VA: Insurance Institute for Highway Safety.
15. Kidd, D.G.; Cicchino, J.B.; Reagan; I.J.; and Kerfoot, L.B. In press. Driver trust in five driver assistance technologies following real-world use in four production vehicles. *Traffic Injury Prevention.*
16. Highway Loss Data Institute. 2015. Predicted availability of safety features on registered vehicles — a 2015 update. *HLDI Bulletin* 32(16). Arlington, VA.
17. Cicchino, J. B.; Zuby, D.S. In press. Prevalence of driver physical factors leading to unintentional lane departure crashes. *Traffic Injury Prevention.*
18. Insurance Institute for Highway Safety. 2016. Special issue: autonomous vehicles. *Status Report* 51(8). Arlington, VA.

Mr. LATTA. Well, thank you very much. And Dr. Stepper, you are recognized for 5 minutes for your opening remarks. Thank you very much for being here.

STATEMENT OF KAY STEPPER

Dr. STEPPER. Thank you Chairman Latta, Ranking Member Schakowsky, and members of the committee for the opportunity to testify before you today. My name is Kay Stepper, vice president with responsibility for the Driver Assistance and Automated Driving Systems for Bosch in the United States. At Bosch we are proud to be inventive for life, and I am honored to discuss an issue that is one of the pillars of our everyday work at Bosch: to save lives.

Bosch has a long history in the United States. Robert Bosch himself established the first office in the United States in New York City in 1906. Now in 2017, Bosch companies operate more than 100 sites across the country. Bosch believes that automated driving is the future of mobility, and leading the way to safe, agile, and automated driving is our guiding principle. Worldwide, Bosch has more than 2,500 engineers and researchers working on the topics of automated driving and advanced driver assistance in our autonomous driving tests that is conducted in the United States, Germany, Japan, and Australia.

Preliminary 2016 data from the National Safety Council projects that as many as 40,000 people died in motor vehicle crashes last year. The magnitude of the safety crisis is such that we must seek active means to increase deployment of technologies that can support drivers and reduce accidents and injury rates. Driver assistance systems such as automatic emergency braking and blind spot detection can assist in reducing the rising fatality and injury numbers that we are facing in the United States today.

In the near term, it is critical that Government and industry continue to work together to help increase consumer access to and understanding of these advanced technologies. I commend the committee for calling this hearing and for focusing its attention on two topics that lie at the heart of this transformation in vehicle mobility: the levels of automation and the importance of the deployment of driver assistance systems as a foundation for automated driving.

Unfortunately, these topics are often overlooked in the overall dialogue about automated driving. The truth is that many drivers and passengers are already experiencing the benefits of vehicle automation every single day. The active safety system electronic stability control is integrated into every new light-duty vehicle sold in the United States today. This revolutionary technology invented by Bosch engineer Dr. Anton van Zanten has saved thousands of lives. A 2014 report from NHTSA found that ESC saved close to 4,000 lives during the 5-year period from 2008 to 2012.

Automated driving will bring great benefits and pave the paths forward a new vision of personal and collective transportation. However, it will take time to achieve fully automated driving and it will be an evolutionary process, building up on the stepping stones of active safety, driver assistance, and crash avoidance system.

In discussing the evolution toward automated driving I want to emphasis that Bosch strongly supports NHTSA's decision to adopt the SAE J3016 framework for levels of automation as part of the Federal Automated Vehicle Policy. This is a major step toward harmonizing and establishing a common set of definitions across the various stakeholders involved in these efforts. Bosch wishes to highlight automatic emergency braking as one clear example of how drivers are being introduced to automation in a gradual manner, and also of how automation intervention by the vehicle can provide the greatest benefit in terms of accident reduction.

Suppliers play an important role in the innovation cycle and many suppliers such as Bosch conduct extensive testing in the lab on test tracks and on public roads. Suppliers presently face several obstacles in carrying out this testing on public roads, and we respectfully request that the committee consider extending the FAST Act exemption to include suppliers with active and established research and development programs in the United States.

Bosch position on the need for improved consumer education is well known. We have urged NHTSA and the U.S. Department of Transportation for many years to include crash avoidance system as a key component of the vehicle 5–Star rating and to provide additional information to consumers through the Monroney label. Bosch believes that displaying crash avoidance systems as part of the official safety portion of the Monroney label and particular in the form of 5–Star rating, as the most effective means to help driver consumer awareness and eventually consumer demand for such technologies. Without the clear presence of crash avoidance and mitigation technologies on the most recognizable feature for consumers, the physical Monroney label as affixed to the vehicle, consumer education will continue to lag.

The adaption of crash avoidance technologies into NCAP would be a very significant improvement and one which we believe will bring about immediate benefits as well as paving the path toward the attainment of automated driving in the future. Bosch encourages Congress and NHTSA to cooperate a path forward for the U.S. NCAP to become an effective means of encouraging the enhanced adoption of these lifesaving systems. Bosch truly believes that a 5–Star rating is the most effective means to translate the presence and performance of crash avoidance technologies into an easy-to-understand indicator for consumers.

Thank you again for the opportunity to speak before the committee. I welcome any questions you may have.

[The prepared statement of Dr. Stepper follows:]

Testimony of Dr. Kay Stepper

Vice President for Automated Driving and Driver Assistance Systems

Chassis Systems Control

Robert Bosch LLC

Hearing on

Self-Driving Cars: Levels of Automation

House Energy & Commerce Committee

Subcommittee on Digital Commerce and Consumer Protection

United State House of Representatives

March 28, 2017

Robert Bosch LLC
Testimony before the House Energy and Commerce
Subcommittee on Digital Commerce and Consumer Protection
March 28, 2017

Background

As a global Tier One supplier, Bosch is working diligently to make Automated Driving a reality. We currently employ more than 2,500 engineers working worldwide on the topics of automated driving and advanced driver assistance.

Key Areas

Automated Driving and Innovation: Continued federal collaboration and support for research and testing is vital to keeping the U.S. and the automotive industry at the forefront of technological innovation. Lawmakers and regulators should permit the safe and responsible testing of advanced safety and automated technologies on public roads. The Federal Automated Vehicles Policy, issued by NHTSA in 2016, represents an important step forward but critical issues must still be addressed.

Driver Assistance / Crash Avoidance Systems: The automotive industry continues to develop and bring to market innovative safety technologies that have made a significant difference in reducing fatality and injury rates. Adoption rates, however, remain low and additional actions must be taken to encourage the installation of these technologies.

Consumer Education and NCAP: The U.S. New Car Assessment Program (NCAP) is the leading mechanism through which the federal government communicates vehicle safety information to consumers. Although the NHTSA has proposed an update to NCAP, the current NCAP model, which is focused solely on crashworthiness and rollover propensity, is outdated and should be modernized.

Chairman Latta, Ranking Member Schakowsky, members of the Committee, thank you for the opportunity to testify before you today.

My name is Kay Stepper, Vice President with responsibility for the Driver Assistance and Automated Driving Systems for Bosch in the United States. At Bosch we are proud to be "Invented for Life" and I am honored to discuss an issue that is one of the pillars of our everyday work at Bosch: to save lives.

Robert Bosch founded the company in 1886, when he opened the "Workshop for Precision Mechanics and Electrical Engineering" in Stuttgart, Germany. Today, the Bosch group of companies employ more than 390,000 associates around the globe, including nearly 18,000 in the United States.

Bosch has a long history in the United States. In fact, the U.S. played a pivotal role in the history of Robert Bosch himself. At the age of just 23, he ventured across the Atlantic, traveling to the U.S. to work with Edison and gain insights into electrical engineering. He subsequently established an office in New York City in 1906. Now, in 2017, Bosch companies operate more than 100 manufacturing, development, sales, service and administrative sites across the country with a significant presence in Michigan, South Carolina, Illinois, California, Wisconsin, and Kentucky. We also have three dedicated Research and Development Centers in the U.S.; they are located in Pittsburgh, PA; Cambridge, MA and Palo Alto, CA.

Bosch has four business sectors – Mobility Solutions, Energy and Building Technology; Consumer Goods; and Industrial Technology. Mobility solutions is our largest sector, comprising approximately 60 percent of our business and representing 217,000 associates.

Bosch is very active at every level of autonomous driving. As a Tier One full systems supplier, Bosch understands the entire automated driving system from requirements derivation to turn-key solutions. Ranging from individual components such as sensors, electronic control units, brake systems, steering, to the overall system, we develop and supply almost every element required for automated driving. Bosch is the world's largest manufacturer of MEMS and radar sensors and a leading global manufacturer of mono- and stereo-vision cameras, ultrasonic sensors, braking and steering systems. With this broad product reach, combined with our expertise in cybersecurity protection, Bosch is uniquely positioned to help drive the creation of a full system approach for our customers.

Bosch is advancing artificial intelligence. At the Bosch Connected World 2017 conference in Berlin, Bosch presented an onboard computer for automated vehicles. Thanks to artificial intelligence (AI), the computer can apply machine learning methods. The AI onboard computer is expected to guide self-driving cars through even complex traffic situations, or ones that are new to the car.

Bosch believes that automated driving is the future of mobility and "Leading the way to safe, agile and automated driving" is our guiding principle. Bosch has more than 2,500 engineers working worldwide on the topics of automated driving and advanced driver assistance in order to achieve this goal and our autonomous driving testing is conducted in the U.S., Germany, Japan, and Australia.

Accident statistics indicate that more than 90 percent of all crashes are caused by human error. A forward-thinking vehicle which takes over dedicated driving tasks could make the vision of injury and collision-free driving a reality. NHTSA's preliminary numbers for the first nine months of 2016 show that an estimated 27,875 people died in crashes – an eight percent increase over the first 9 months of 2015.[1] Preliminary 2016 data from the National Safety Council projects that as many as 40,000 people died in motor vehicle crashes last year.[2]

The magnitude of this safety crisis is such that we must seek active means to increase deployment of technologies that can support drivers and reduce accident and injury rates. Driver assistance systems such as Automatic Emergency Braking (AEB) and Blind Spot Detection (BSD) can assist in reducing the rising fatality and injury numbers that we are facing in the United States today. In the near term, it is critical that government and industry continue to work together to help increase consumer access to and understanding of these advanced technologies.

In January 2017, Bosch released a study "Connected Car Effect 2025"[3] which investigated what mobility technology will mean specifically for the US, Germany and the major cities of China. The result: safety systems and cloud-based functions can prevent around 260,000 injury accidents, save 390,000 tons of CO_2 emissions and offer drivers many hours of more time for other activities. Over 260,000 accidents involving personal injuries (US: 210,000, China: 20,000, Germany: 30,000) will be avoided annually – as many accidents as occur within two years in Germany's capital city of Berlin. The Study predicted that 350,000 fewer people would be injured by traffic accidents – the same as 12 years without traffic injuries in Los Angeles. In the US alone, there will be 290,000 fewer (China: 25,000, Germany: 37,000).

I commend the Committee for calling this hearing and for focusing its attention on two topics that lie at the heart of this transformation in

---

[1] DOT HS 812 358: A Brief Statistical Summary - Early Estimate of Motor Vehicle Traffic Fatalities For the First 9 Months of 2016; January 2017
[2] National Safety Council Press Release, February 15, 2017, "Motor Vehicle Deaths in 2016 Estimated to be Highest in Nine Years"
[3] "Connected Car Effect 2025" conducted by Bosch and the consulting firm Prognos, January 2017

vehicle mobility: the levels of automation and the importance of the deployment of driver assistance systems as a foundation for automated driving. Unfortunately, these topics are often overlooked within the overall dialogue about Automated Driving. The truth is that many drivers and passengers are already experiencing the benefits of vehicle automation every day. The active safety system Electronic Stability Control (ESC) is integrated into every new passenger car sold in the United States. This revolutionary technology, invented by Bosch engineer Dr. Anton van Zanten, has saved thousands of lives. A 2014 report from NHTSA found that ESC saved close to 4,000 lives during the 5-year period from 2008 to 2012[4]. The technology works by monitoring driver intent and vehicle direction and by automatically applying braking force as needed to prevent a loss of control. Most drivers are not even aware of its support. This intervention is communicated to the driver as a mere flash of the indicator light on the dash, but the real world result is often a life saved or a serious injury mitigated.

Automated driving will bring great benefits and pave the path toward a new vision of personal and collective transportation. However, it will take time to achieve fully automated driving and it will be an evolutionary process, building upon the stepping stones of active safety, driver assistance and crash avoidance systems. The first wave is already here in the form of driver assistance systems that utilize automation to increase safety. The next phase will consist of partially-automated functions, such as traffic jam assist, which are available in the market but not deployed in great numbers.

In discussing the evolution toward Automated Driving, I want to emphasize that Bosch strongly supports NHTSA's decision to adopt the

---

[4] DOT HS 812 042, June 2014, Estimating Lives Saved by Electronic Stability Control, 2008–2012

SAE International (SAE) J3016 framework for levels of automation as part of the Federal Automated Vehicles Policy. This is a major step toward harmonizing and establishing a common set of definitions across the various stakeholders involved in these efforts. Without a common taxonomy and understanding of the different levels of automation, it will be considerably more difficult to make the necessary strides toward full automation. In fact, the lack of a common "language" and standardized descriptions is one of the obstacles that has hindered the understanding and adoption of Advanced Driver Assistance Systems (ADAS).

A 2015 study conducted by the Boston Consulting Group (BCG), on behalf of the Motor & Equipment Manufacturers Association (MEMA), determined that the widespread installation of ADAS technologies could prevent about 9,900 fatalities each year and save more than $250 billion annually in societal costs[5] in the United States. The BCG found that, at that time, ADAS features were not present in a high number of vehicles and that their share of the U.S. market was growing at only 2 to 5 percent annually.

Bosch wishes to highlight Automatic Emergency Braking as one clear example of how drivers are being introduced to automation in a gradual manner and also of how automation and intervention by the vehicle can provide the greatest benefit in terms of accident reduction. The full suite of AEB, also known as Crash Imminent Braking, consists of three technologies. The first is Forward Collision Warning which simply alerts the driver to the fact that he/she is getting very close to the vehicle in front of them. The next stage technology, termed by NHTSA as Dynamic Brake Support, actually prepares the brakes and pre-fills them so that the driver will immediately have full braking power when he/she engages the brakes

---

[5] A Roadmap to Safer Driving Through Advanced Driver Assistance Systems (ADAS), Boston Consulting Group, September 2015

to slow or stop the vehicle. The driver does not feel a demonstrable difference in the brake but he/she receives enhanced braking power to reduce the stopping distance. AEB is the last technology in this cascade. If the driver takes no action, then the system engages the brakes on its own and stops the vehicle to prevent or mitigate the crash.

These types of crashes remain a leading safety concern in the United States. Bosch's internal analysis of NHTSA's 2013 NASS[6] data indicated that approximately 33 percent of collisions with injuries and fatalities were rear-end crashes. Most drivers believe that they are observant and fully aware of their surroundings, but Bosch's research found that drivers often fail to detect the obstacle in rear end crashes. In cases where the driver did detect the obstacle, approximately 49 percent of the drivers failed to apply adequate braking force in order to avoid the collision. Further, 31 percent of drivers failed to even apply the brakes at all. This cascade approach optimizes the system's ability to support and assist the driver. It exemplifies the manner in which increasing levels of automation can help to supplement the driver's own abilities.

Bosch, together with its customers and other suppliers, has also devoted considerable resources to tackling the growing safety problem of pedestrian fatalities and injuries. Technology, in the form of advanced pedestrian detection and braking technologies, offers us the opportunity to notably mitigate and, in some cases, prevent crashes involving vulnerable road users. In 2015 there were 5,376 pedestrians killed in traffic crashes in the U.S., a 9.5 percent increase from the 4,910 pedestrian fatalities in 2014. This figure represents the highest number of pedestrians killed annually since 1996. On average, a pedestrian was killed nearly every 1.6 hours and injured more than every 7.5 minutes in traffic crashes in 2015.[7]

---

[6] National Highway Traffic Safety Administration, National Automotive Sampling System (NASS)
[7] DOT HS 812 375, NHTSA Traffic Safety Facts, February 2017

Bosch has developed forward pedestrian Automatic Emergency Braking systems as well as rear Automatic Braking systems to address these types of crashes. We are applying the same strategies utilized in AEB, cascading from a driver warning to a full automatic intervention. Also, as mentioned before, Bosch has been developing Artificial Intelligence for use in vehicle automation. Bosch's AI onboard computer can recognize pedestrians or cyclists.

Suppliers play an important role in the innovation cycle and many established suppliers, such as Bosch, conduct extensive testing in the lab, on test tracks, and on public roads. These activities are integral to the development and maturity of the technology needed for automated driving. Our engineers conduct extensive track testing and simulation work; but nothing can replace the importance of on the road testing and validation. In addition, these efforts are intrinsic to enabling suppliers to develop their own robust and comprehensive offerings for OEMs and support the competitive challenge to deliver the most effective and cost-sensitive software options for automakers. Prohibitions and delays that impede on-road testing will slow this process at the supplier level and; thereby, inhibit the overall progression of automated driving technology. For Bosch, reliability and robustness are the top priorities when it comes to safety systems. This requires the use of thoroughly tested and approved software. Bosch emphasizes that these development vehicles are driven exclusively by trained test drivers at Bosch and equipped with special safety concepts to enable the driver to reassert control at any time. After a successful release procedure on test tracks, we take the system on public roads to conduct evaluations in a real environment but always under supervision of a trained driver accompanied by a test engineer monitoring the system.

Suppliers presently face several obstacles in carrying out this testing on public roads and we respectfully request that the Committee consider extending the FAST Act exemption to include suppliers with active and established research and development programs in the U.S.

Bosch has been a passionate advocate for the deployment of driver assistance systems. We continue to view these technologies as the most effective and immediate means to reducing fatalities and injuries. We have worked diligently as a company to make these systems more accessible to all consumers. By developing cost effective components, such as Bosch's mid-range radar sensor, we have sought to support the distribution of these systems to all makes and models of passenger vehicles. Series production of Bosch radar sensors began in 2000. In 2016, Bosch delivered its ten-millionth radar sensor.

Bosch's position on the need for improved consumer education is well known. We have urged NHTSA and the U.S. Department of Transportation for many years to include crash avoidance systems as a key component of the vehicle 5-star rating and to provide additional information to consumers through the Monroney Label. Bosch believes that displaying crash avoidance systems as part of the official safety portion of the Monroney Label, and particularly in the form a five star rating, is the most effective means to help drive consumer awareness and eventually consumer demand for such technologies. Without the clear presence of crash avoidance and mitigation technologies on the most recognizable feature for consumers - the physical Monroney Label as affixed to the vehicle – consumer education will continue to lag.

Support for the proposed inclusion of crash avoidance technologies was also confirmed by prominent groups such as the National Safety Council and the Insurance Institute for Highway Safety in their formal responses to the NHTSA proposed NCAP update, issued in December 2015.

The adoption of crash avoidance technologies into NCAP would be a very significant improvement and one which we believe will help bring about immediate benefits, as well as paving the path toward the attainment of automated driving in the future. Based on Bosch's analysis of the 2013 NASS data, crash avoidance technologies such as forward collision warning, automatic emergency braking, lane departure warning, lane keeping systems, blind spot detection, lane change assist and pedestrian crash avoidance systems have the potential to avoid or mitigate up to 64 percent of passenger car and light-duty truck collisions resulting in injuries and fatalities in the United States.

As part of the proposed NCAP update issued in December 2015, NHTSA had proposed separate ratings for crash avoidance and pedestrian protection, as well as a combined overall vehicle rating. Bosch acknowledges that there were many issues that still need to be addressed and fleshed out as part of the proposals and we are aware that many entities raised legitimate concerns relative to the proposed changes to the crashworthiness section of the proposal. Our intent is simply to encourage Congress and NHTSA to cooperate and find a path forward for the U.S. NCAP to become an effective means of encouraging the enhanced adoption of these life-saving systems. Bosch truly believes that a five star rating is the most effective means to translate the presence and performance of crash avoidance technologies into an easy-to-understand indicator for consumers.

Thank you again for the opportunity to speak before the Committee. I welcome any questions you may have.

Mr. LATTA. Well, thank you very much for your testimony and that will conclude our opening statements from our witnesses. Again we appreciate you being here, and I will begin the questions if I may.

And if I could, Mr. Zuby, I would like to just follow up what you said what you said. A lot of the drivers out there driving the vehicles that have a lot of this technology are not using it. Is it because, you know, is it too difficult for them to understand maybe from reading the instructions in the manual or they just don't want to bother with doing it, or what are you finding out there why people aren't using that technology?

Mr. ZUBY. Right. So we think that one of the reasons that people aren't using lane departure warning technology is because they find it annoying. The way that technology works today is that it basically gives you a warning which may be an audible beeping or a vibrating of the steering wheel or vibrating of the seat when you transgress a lane line without signaling your intention to do so.

So one way to think of the current technology is it is sort of a turn signal nanny rather than warning the driver about an imminent danger. And when we interview, or rather survey drivers with the technology that is one of the things that they tell us is the lane departure warning is very annoying. Systems that interact with the driver less frequently like front crash prevention are much more likely to be left turned on. In the studies that we have done we find that AEB and front crash warnings are left on in 90 percent or more of the vehicles, whereas we only see about 50 percent of lane departure systems left on.

The other thing that our research is finding is that the design of the lane departure warning seems to have an influence. So people don't like the audible alerts, but when the system alerts them about crossing the lane line with a vibrating steering wheel or a vibrating seat they are much more likely to leave it on. And we also find that if the car takes some steering action in response to, you know, transgressing the line that too leads to higher use rates than the original systems which only warned the driver with an audible warning.

Mr. LATTA. Well, thank you very much.

Mr. Klei, if I could ask you a little bit about especially on the cyber side, in your testimony you mentioned how driver assistance systems will require sensors to gather data about a vehicle's surrounding environment in order to adequately assist that driver. How is Continental thinking about the privacy and security of the advanced driver assistance systems and crash avoidance systems, and what is Continental doing to secure those systems against cyber threats?

Mr. KLEI. Thanks for the question, Chairman Latta, and it is a great question and it is something that at Continental we have been thinking about for many years. Cybersecurity is not new with automated driving or the advanced driver assistance systems. It has been a discussion point and a key development area for us for many, many years ever since, really, electronics started to come into the car.

I would say the connection to the cloud, the connection with all the 4G connections that are now available open up a new opportunity for those cybersecurity threats. We have developed an entire competency center in our company that is used extensively for cybersecurity and we are trying to install all the different protections that we can from known cybersecurity attacks.

But many people say should we have a cybersecurity specification it is dynamic. Every day there is new threats. Every day there is new opportunities that emerge. So we have to work together with our OEM partners, suppliers, and the Government to look at ways we can work together to identify and eliminate those cybersecurity attacks. But we clearly have a competency center, we think very much about it, and it is clearly a challenge as we bring many of these technologies into market. But it is not new. It has been thought about and developed for many, many years.

Mr. LATTA. Well, if I could also, Dr. Stepper, would you like to comment on that on what Bosch is doing in this area on the cyber side?

Dr. STEPPER. Yes. Thank you, Chairman, for the question. Bosch has been very active on this topic for cybersecurity protection. We believe very much in a layered approach, layered in a sense that there is hardware layer, software layers, and architectural layers that need to be introduced. We actually established a center of competency for cybersecurity back in 2010, and we already established additional units within Bosch that work specifically on software solution to help our OEM partners to protect against cybersecurity threats.

Mr. LATTA. Well, thank you very much.

And also, Dr. Klei, could I ask a real quick question because my time is running out here, commenting on SAE levels of automation and why they are important to the industry standard of fully self-driving cars.

Mr. KLEI. Certainly we very much support the adoption of the SAE standards. We think a standard that clearly defines what the levels of automation are, are very useful as we start to develop and deploy these technologies. The consumers are often confused by the various naming and the various levels. And I think we as an industry have a lot of work to do to improve that communication and education of the consumers.

Suppliers have a role in this. The OEMs have probably the largest role because they are the ultimate touch point with consumers. And then of course any assistance from the Government and other outside agencies are very, very beneficial. So we very much support it and we think everyone has a role in educating so that the naming of these technologies really describe what it can do and people don't get confused.

Mr. LATTA. Well, thank you very much. And my time is expired and I will now recognize the gentlelady from Illinois, the ranking member of the subcommittee, for 5 minutes.

Ms. SCHAKOWSKY. Thank you, Mr. Chairman.

Dr. Stepper, your testimony mentions rear automatic emergency braking systems and I am wondering if you could discuss how that could help prevent backover accidents.

Dr. STEPPER. Yes. Thank you, Ranking Member Schakowsky, for the question. The rear automated emergency braking is a relatively recent addition to the automatic emergency brake suite of functions that we have. We already have a mandate in the United States starting in 2018 for backover legislation to have a rearview camera installed in each and every vehicle.

So we have already a basis of the technology in there, and we also see that especially with pedestrian incidents that we see in rear backover situations this technology could really help not only to protect from material damage but saves lives and prevent injuries.

Ms. SCHAKOWSKY. And is this feature available today in any makes or models?

Dr. STEPPER. It is available today but still in very, very small numbers. There are a few select vehicles in the United States today sold with this. The installation rate overall is less than five percent, in contrast to forward-looking automatic emergency braking where you look more between a 20 to 25 percent installation rate today already.

Ms. SCHAKOWSKY. You also mentioned pedestrian automatic emergency braking. Is that any different from AEB when another car is in front of the vehicle?

Dr. STEPPER. It is another progression and another step in the full AEB suite. The automatic emergency braking for vehicles was invented first and brought to market. Pedestrian automatic emergency braking has a little bit of a different requirement in the sense that you need to have a very wide field of view to recognize crossing pedestrians and not only at higher speeds, but especially in urban scenarios at lower speeds. So and therefore it is different in the sense that the requirements on the technology are different and it is already part of Euro NCAP in the European Union as a requirement moving forward.

Ms. SCHAKOWSKY. Thank you. Mr. Zuby, I wonder if you have looked into these technologies and if you have any comments on that.

Mr. ZUBY. Yes. We have been looking into these technologies and we have worked up a series of tests that we intend to start using to promote the idea of reversing automatic braking. We think that that may be an additional thing that is needed to address backover crashes because the experiments that have been run using cameras show that while they definitely improve the situation and help drivers avoid running into things that are behind their vehicle that they don't expect to be behind their vehicle, they are not a hundred percent effective because the driver needs to be looking at the camera at the same time that the person or object behind them is in the view of the camera.

So automatic braking, I think, can augment the benefits that we get from the technology looking rearward in the camera during reversing maneuvers. We are also looking at pedestrian—by the way, my guys have identified, I think, 14 models of cars sold in the current model year that are equipped with reversing AEB. We are also looking at pedestrian detection. And it is a slightly more difficult problem for the technology to solve because of the field-of-view issue and the fact that pedestrians can change direction and change their movement very quickly.

Ms. SCHAKOWSKY. Because I have been so involved in the issue of the cameras and you say it is not a hundred percent, have you estimated how effective it is or how many times it does fail to prevent an accident?

Mr. ZUBY. Well, so in experiments we find that it reduces the likelihood that you are going to back over something that is in your path by about two thirds.

Ms. SCHAKOWSKY. OK. So you have done years of research on AEB systems. Can you give us more details on how these systems work and why they save lives?

Mr. ZUBY. So the current AEB systems mainly prevent front-to-rear crashes. They are effective at preventing those kinds of crashes, and even when they don't prevent the crash they reduce the risk of injury. Front-to-rear crashes don't result in a lot of fatalities. It is in the neighborhood of about 800, 900 people a year out of the nearly 40,000 die in front-to-rear crashes. So even if a technology were to prevent all of the rear crashes, it would have a small dent on fatalities.

But the sensors that are needed for AEB are sensors that will be needed to address other types of crashes, you know, leverage the technology to address other kinds of crashes that do account for more fatalities.

Ms. SCHAKOWSKY. Thank you. I see I am out of time, I yield back. Thank you.

Mr. LATTA. Well, thank you very much. The gentlelady yields back, and the Chair now recognizes the gentleman from Illinois for 5 minutes.

Mr. KINZINGER. Thank you, Mr. Chairman. Thank you all for being here and taking some time with us today. It is an important hearing on the future of self-driving cars and specifically the opportunity to learn more about the advanced driver assistance systems that is saving lives today and it is also paving the way to fully autonomous vehicles.

Dr. Stepper, in your testimony you highlighted the importance of the SAE framework for the various stakeholders in autonomous vehicles and the lack of common language for advanced driver assistance systems. How has this lack of a voluntary standard impacted Bosch's ability to bring technology to the market?

Dr. STEPPER. Thank you for the question, Congressman. Very clearly, the lack of clear language and common taxonomy has resulted in some confusion at the consumer side: What is really my car doing with the different technologies that we have? So as Mr. Gouse has very graphically illustrated in his chart, there is well defined levels of 0 to 5 for automation, and coupled with a very active consumer education campaign we can really educate consumers what they can expect.

Is it just a warning that my vehicle will provide or is it actually an actual intervention like an active braking situation or can I take my hands and my feet off the controls and the car will drive by itself? And what we have found clearly is that the lack of such common language really has led to confusion on the consumer end, and we really commend the National Safety Council together with the University of Iowa joining the Road to Zero campaign and actually establishing a Web site that is called mycardoeswhat.org to educate consumers of what is actually in their vehicles today because it can be so confusing.

Mr. KINZINGER. We should do a my-congressman-does-what. Mr. Gouse, what are the challenges to adopting a voluntary consensus standard and what efforts are underway to provide a common language for advanced driver assistance systems?

Mr. GOUSE. Thank you, Congressman, for the question. It is an emotional question internally, because it is very difficult to raise awareness that our documents even exist to a variety of stakeholders that don't traditionally know that they even use this. We were working with the American Association of Motor Vehicle Administrators and they didn't even know that the license plate geometry was our standard. So that was our beginning point. And we told them we had this document in works at the same time NHTSA had their levels of automation in works, and with differing vocabulary and differing levels it confused the issue a lot. Fortunately, NHTSA decided to adopt the SAE language, and then AAMVA and through the States that proliferated. That is one example.

The same thing is happening all over the world. For the driver assistance systems, same thing, we have a standard that is called Active Safety Systems Terms and Definitions. It is a fairly easy read. It is not really riveting like a novel, but it is a fairly easy read and we are trying to get that language adopted too. And as you hear today, we even use different terms ourselves and I agree it is confusing.

Mr. KINZINGER. Let me add on. Are there any policies, developing policies that you are concerned with as you are seeing them right now?

Mr. GOUSE. The States that are unaware or choosing not to use a common terminology and the common taxonomy, I believe, will result in a patchwork of very difficult to understand and operate in environments. This is happening now at the testing level where they are passing regulations permitting testing of various levels of automation in nonsalable vehicles. So it is a concern.

Mr. KINZINGER. And then we will go with Dr. Stepper on this one. When you look at educating the public about the benefits and the limitations of various systems, especially for systems like automatic emergency braking that provides a lot of value to the customer, but the customer, the consumer may not be aware that the technology is assisting the driver. Mr. Zuby mentioned that lane maintenance systems were only turned on in 51 percent of the vehicles that IHS observed. How do your companies, how does your company work with the consumers to build confidence in the technology so it is being fully utilized?

Dr. STEPPER. So thank you for the question, Congressman. Clearly we work with activities like the Road to Zero and the activities from the National Safety Council as well as the University of Iowa. We work very closely with our OEM customers, for example, in joined co-marketing campaigns to educate dealers, because at the end of the day new vehicles are being bought from dealerships and consumers are being consulted by dealership personnel and that is really your first touch point of a new vehicle purchase and understanding of what this vehicle really has on board in terms of technology.

So we work very actively with several OEM customers on this topic to make tours to make joint marketing campaigns around the country to educate dealerships on this topic so they can explain what is installed on the vehicle. Again I want to emphasize an additional mention of these crash avoidance technologies. In a 5–Star rating, incorporating crash avoidance technologies could also very much help in that regard because now the dealership personnel would have the Monroney label right in front of them to help them guide the consumer through the purchase.

Mr. KINZINGER. Thank you. And I have some more questions; I will submit them for the record. Mr. Chairman, I yield back.

Mr. LATTA. Thank you very much. The gentleman yields back, and the Chair now recognizes the gentlelady from Michigan for 5 minutes.

Mrs. DINGELL. Thank you for the recognition, Mr. Chairman, and for your continued interest in the automated vehicles. As you all know it is a subject that I really care a great deal about. As I stated in the last hearing on this issue, I believe it is critical that the Congress, the administration, the industry, and safety advocates all come together on a common framework for automated vehicles. Too much is at stake and we have got to get it right.

Legislation will be needed to facilitate the deployment of higher level automated vehicles, and I support raising the statutory exemption caps as an interim solution while directing NHTSA to amend existing vehicle safety standards as they relate to human operated controls. And I think a lot of people don't understand what some of the regulations are because they have been there for so long.

Great strides in vehicle automation are being made. Proud of it that a lot of it is in my district in developing safety technologies that have the potential to reduce roadway deaths, and I believe helping them get to market could have a significant impact on public safety, and I have got some questions to help the committee examine these issues.

My first questions are for Mr. Gouse of SAE, and if you could just do yes or no, please. Is it correct that SAE Levels 0 to 2 contemplate that a human driver will perform all or some aspects of what is known as the dynamic driving task?

Mr. GOUSE. Yes.

Mrs. DINGELL. Is it correct that SAE Level 3 contemplates that a human driver must be in the loop and prepared to respond to a request by the vehicle to take over the dynamic driving task?

Mr. GOUSE. Yes.

Mrs. DINGELL. Now is it true that an SAE Level 4 vehicle is one that is capable of performing all aspects of the dynamic driving task in a given situation also known as the operational design domain?

Mr. GOUSE. Yes.

Mrs. DINGELL. And a Level 5 vehicle can handle all aspects of driving under all conditions?

Mr. GOUSE. Yes.

Mrs. DINGELL. Thank you. Now these questions are for all four witnesses. Is it true that companies like FCA, Ford, and GM in Michigan are developing and currently deploying SAE Levels 1 and 2 systems? Anyone can say yes or no.

[Chorus of yeses.]

Mrs. DINGELL. Thank you. Is it true that these traditional automakers and others like Waymo are developing Level 4 systems at the same time?

Mr. KLEI. Yes.

Mrs. DINGELL. In other words, these companies aren't necessarily pursuing a sequential progression through the SAE Levels to full vehicle automation; is that correct?

Mr. KLEI. No.

Mrs. DINGELL. That is not correct. So you think they are going 1, 2, 3, 4 or are they going from 2 to 4?

Dr. STEPPER. If I may jump on this one, Congresswoman Dingell, it depends on the automaker. Some absolutely proceed along the path, Level 0, 1, 2, 3, 4, and 5; some other ones may skip Level 3. There is no common answer. But some of them that you mentioned are indeed following exactly along the path of what Mr. Gouse has presented.

Mrs. DINGELL. And others are skipping. Is it true that a number of existing NHTSA safety standards require human operation of vehicle controls that may not be necessary if there is no human driver, such in Level 4 or 5?

Mr. ZUBY. Yes.

Mr. GOUSE. Yes.

Mrs. DINGELL. Do you all have good—I don't know if I am—my staff wants me to keep moving. But I think people don't know that a NHTSA requirement requires a foot on a brake and it is not necessary at 4 or 5, so——

Dr. STEPPER. That is correct.

Mrs. DINGELL. Thank you. Should NHTSA amend existing safety standards to clarify how they apply to higher level automated vehicles without drivers?

[Chorus of yeses.]

Mrs. DINGELL. Do all of you agree on that?

Mr. GOUSE. Yes.

Mrs. DINGELL. Well, I am running out of time, so I am going to—I have lots of questions but—and, for the record, I may submit some more, Mr. Chairman. But I want to commend the chairman for holding this important hearing to help educate members on the issues because it is really important that we get it right. Automated vehicles are going to be developed and they are going to be developed internationally if we don't take the lead on making sure we do it, develop them here and that these technologies are developed in the United States of America. So I look forward to working with my colleagues on both sides of the aisle in a bipartisan manner to achieve this goal.

Thank you all for being here today. Thank you, Mr. Chairman. I yield back my 15 seconds.

Mr. LATTA. Thank you very much. The gentlelady yields back the balance of her time and the Chair now recognizes for 5 minutes the gentleman from Mississippi, the vice chairman of the subcommittee.

Mr. HARPER. Thank you, Mr. Chairman. And again thanks to each of you. This is, you know, it is just mind boggling the possibilities and we have just barely scratched the surface. And, you know, I can't imagine what it will be like we come back in 5 years and just discuss what we are doing next. I mean this is really remarkable. So thanks for the involvement that each of you and each of your companies have.

And Mr. Klei, thank you very much. We are excited about the presence of the new Continental Tires facility that will be opening in Mississippi. I think that was a great decision. We are honored to have a part of your company that will be there, and I wanted to talk to you for just a minute.

Obviously, the intellectual disabilities issue is important. My wife and I have a son who is 27 years old who has Fragile X syndrome. He graduated from a special program at Mississippi State University. He works Monday through Friday. My wife has to drive him every day and drop him off and pick him up. So it is something for many families, this is an important issue. So are advanced driver assistance systems at a point where they are able to provide new transportation opportunities to the disabled community?

Mr. KLEI. Certainly is it an important topic, and thank you for the question, Congressman. It is something that I think, as an industry we are working very hard, and it is not just for the automated driving technologies in general. We are trying to make mobility more available and safer for all, and I think the advancements in automated driving are clearly going to move that forward.

Are they ready today to take over all driving tasks for someone that can't drive today? Not necessarily; over time, absolutely. We believe when we get to Level 4 and Level 5, absolutely it is going to provide mobility for many people that today don't have that mobility. The Waymo development, their first example that they showed was someone that was blind. And that is a huge statement for the potential mobility promise for the elderly, the blind, and every disabled person in the United States will have mobility, and it is an important step for them, but also for society.

Mr. HARPER. Well, we are excited that Continental is taking that into consideration in the development of this.

Dr. Stepper, will you also comment on that as well?

Dr. STEPPER. Yes. Thank you for the question, Congressman. We are actually working very, very intensively on the aspect of human factors because as we have learned before, on some of the levels of automation the interaction of the human being is still very, very important and part of the requirement for both SAE all the way to Level 3 as we heard earlier.

So in human factors we have done a number of research for user, human-machine interaction perspective, but we have also worked in augmented reality experiences. And that is a topic I just want to make the comment that we are actually going to show a demonstration of augmented reality for automated driving. It is an upcoming experience here on the Hill as the event that is CES on the Hill on April 5th, where all of you of course are invited to experience some of the human factors aspect and how important it is as part of the automated driving equation.

Mr. HARPER. We are expecting self-driving cars to be at Level 5 tomorrow, when most drivers are not Level 5 drivers. Mr. Klei, what do you think Congress should do to facilitate this development in deployment of advanced driver assistance systems at a point where we can assist and not be, let's say, a roadblock to that development?

Mr. KLEI. Thank you, Congressman, a very important question and one that I think we look at a couple different areas. One is the Federal Automated Vehicles Policy that was issued last September. While we commend the NHTSA organization and all the work that they did we think there is a lot more to do.

First of all, when it comes to that policy it really more talks about deployment rather than development, and we think development is an important part of bringing these technologies to market safely and with real world testing.And only through an improvement in that policy can we get there. For example, the policy requires for every software change or every change that we make we have to submit a new exemption. The time to develop those and the time to get the approvals will significantly delay the implementation of this.

I think the other thing is the model State policy. To have a patchwork of State regulations is clearly hindering our ability to test and develop and ultimately commercially deploy these technologies. So there is two examples. I could go on and on about other examples, but clearly there is opportunity to work closer together between ourselves as suppliers, the OEMs, and the Government to really bring these forward in a safe and effective way.

Mr. HARPER. Thanks to each of you. I yield back.

Mr. LATTA. Well, thank you very much. The gentleman yields back, and the Chair now recognizes the gentlelady from California for 5 minutes.

Ms. MATSUI. Thank you very much, Mr. Chairman, and thank you very much for the witnesses for being here today. As many of you know, the FAST Act mandated that self-driving cars could be introduced into commerce solely for the purposes of testing, but only by companies that had at the time of the law's enactment already manufactured and distributed motor vehicles in the United States.

In addition, legislation has been proposed in some States that would allow only traditional car manufacturers to test and deploy AVs. Some have even speculated that NHTSA's deployment exemptions also could be limited to car manufacturers that already build and distribute motor vehicles in the United States, and I believe we started down this path already.

But Dr. Stepper and Mr. Klei, I know that you have been working with AV components that could benefit from direct testing. What are the barriers to your companies doing testing on your own?

Mr. KLEI. From the Continental side certainly we have talked a little bit about some of those barriers with the ability to test without concern for all the different State regulations. I mean, since the Federal Automated Vehicles Policy came out there has been 48 different bills in 20 States that complicate our development of these technologies. We believe that as suppliers we also need to have the ability to test and develop these. It can't be just the OEMs that in fact do certify vehicles for FMVSS. We as suppliers don't certify vehicles. We develop technologies, we work with our OEM partners to bring them in safely, but we need the ability to develop and test those ourselves, not as a certifying FMVSS body but as one that really looks to develop those.

Ms. MATSUI. Certainly. Dr. Stepper?

Dr. STEPPER. Congresswoman Matsui, thank you for the question. As I mentioned earlier in my testimony, suppliers play a very important role in the innovation cycle. And as a matter of fact, often innovations like electronic stability control, the required sensors like radars, video cameras, ultrasonic sensors, and many of the other active systems, for example, the braking and the steering in the vehicle, is actually coming from the suppliers.

So we do our utmost of course to develop and test and verify these components and systems in the lab with artificial methods like modeling and simulation, but there comes the point where we suppliers need to take these technologies on the road to ensure that they are fully verified and validated before they ever go into consumers' hands. So it is really limiting our ability to test on public roads.

And we understand very clearly that the expansion of the exemption must be handled very carefully and cautiously, but we are very happy to engage actively with the committee on this point.

Ms. MATSUI. Thank you. I understand that different companies are pursuing different strategies in terms of the level of automation in the vehicles they plan to deploy. And as we have been reminded, often it is human drivers that can cause and contribute to accidents with automated vehicles.

Mr. Zuby, are there particular concerns we should consider during a transition when vehicles from all different levels of automation will be on the roads?

Mr. ZUBY. Yes. I think we are already seeing in studying work that Waymo are doing and other automakers that even when the automated cars are driving at a very high level of competency they often are involved in crashes caused by human drivers. And so I think as the testing develop it is important to make sure that there are safeguards that the testing be done in safe ways and not endanger other people and the public, but it will be absolutely necessary to test these things in the real situation because that is where they need to work.

Ms. MATSUI. Right. As companies continue to expand testing of autonomous vehicles, they are all gathering an enormous amount of data about these vehicles. Mr. Gouse, are there any efforts in place to standardize the data that is being collected so that we can learn best practices regardless of where the autonomous vehicles is being tested?

Mr. GOUSE. Ma'am, there are very early efforts going on. You have to understand that it is a very proprietary environment. While these gentlemen are cordial here, they probably want to kill each other sometime over a product.

Ms. MATSUI. I hope not.

Mr. GOUSE. No, no, no. So there are discussions going underway with the associations that they belong to on this and how to collect the data and use it.

Ms. MATSUI. So we are at the very early stages of that right now but it would be very helpful to have the data. So anyway I will yield back my remaining time.

Mr. LATTA. Thank you very much. The gentlelady yields back, and the Chair now recognizes the gentleman from West Virginia for 5 minutes.

Mr. MCKINLEY. Thank you, Mr. Chairman. And last month when we met I said then that I think this is, this whole process is probably inevitable. And as one of just two licensed engineers in Congress, I am intrigued with the problem-solving possibilities that we have with this. I am fascinated with the developments that have occurred so far in lane movement as you referred to it or the braking.

But I am a huge skeptic of driverless cars and I am not buying this one iota yet. I will go with all the others. I can see the possibilities of that. But at the last meeting I raised some questions about IV&V and everyone on the panel had no idea what we were talking about, so I ask you because you are four different people. Are you using IV&V for confirmation of the various steps that we are going through so far?

I am seeing a no all the way around again. If we send a ship to Mars or when we send a satellite into space we run through all the steps to test it for individual verification and validation and make sure that it is going to work because we don't want to rely on competitive peer pressure without having some third party validate what we are doing. And that is what we are looking for, I am going to looking for is third party, because I know companies are going to be under a lot of pressure to skip steps 2 and 3 and go right to 4 if possible or skip 1 and go to 3, whatever that might be they are going to move that because of competitive pressures.

We talked a little bit when one of the things since that time—because I am fascinated with this. Again, it is the engineering. I know this is inevitable. How can we work with this thing to do everything but driverless? So when I have asked the question when I have been back in my district, it is wherever it is we are excited. In fact we are going to have a summit meeting about this, about driverless cars.

But when I have raised the question, Would you put your 6-year-old granddaughter in the car and let her go 40 miles to meet her brother, perhaps, every one of them says no. Now, I know it is going to be evolutionary. They will develop more confidence with it. But when I was hearing about if something goes wrong they are going to transfer operation back over to the person in the car, what happens if it is indeed someone that is intellectually impaired or is inebriated and we have allowed them to get in that car to be able to get home, and then they are turning the transportation over to them when they are doing 60 miles an hour, and they say, "OK, driver, it is your car"?

I have a series of questions about it. I am going to remain a skeptic on this. I want to follow the money. I don't understand other than insurance companies who is really going to benefit for this, but as an engineer let me skip to my last, so ask a question of this. If when we get to steps 4 and 5, because I have designed a lot of bridges, a lot of highways, culverts, I don't know how this is functioning yet, so is there something I should be working in in my old company in engineering that starts to get ready so the cars when we are at steps 4 and 5 there is something, is there a wire in the road, is there something along the guardrail, or is this something merely sensing it? Is this all GPS driven?

I need to have a lot more information before we get anywhere close to that. Because if we are designing all these roads, why aren't we taking those things into consideration now especially with this infrastructure bill that it is going to have? So with that can you tell me what should we be doing in our highways to be ready for steps 4 and 5?

Mr. KLEI. In terms of the highways themselves we have to adapt to the highways, we can't expect the highways to adapt to these systems. That is why real world testing around the world has to happen.

Mr. MCKINLEY. So in that case, Mr. Klei, is it GPS driven or is it sensing the side of the highway?

Mr. KLEI. It is both. It is GPS, it is sensing.

Mr. MCKINLEY. It goes through a tunnel, and in West Virginia, where we have almost 50 percent of the State does not have service, I lose my signal constantly and no one knows where we are. And I don't know what happens at that point, so you are going to have to rely on a lot better control if you are going to use GPS. So if it is going to be sensing how do we do that?

Mr. KLEI. Obviously, the sensory development is a key part of that. But it is not just sensing it is also GPS. It is also vehicle-to-vehicle, vehicle-to-infrastructure, DSRC, all of that coming together will unable that Level 4 and Level 5.

Mr. MCKINLEY. Thank you very much, I have run out of my time. But I want some engineering answers on this, not the 90 percent savings of accidents, because I think it is BS. It is not going to happen, just like we have had the debates here over my 7 years in Congress that, if we stop using coal, we would eliminate 80 percent of the asthma attacks in this country. We know that is false. So I don't want to use a technique or a topic that says we are going to save 90 percent of accidents if we adopt this, I want to have more facts. The engineer in me says I need more facts. So thank you, and I yield back.

Mr. LATTA. Thank you very much. The gentleman yields back and the gentleman from Texas is now recognized for 5 minutes.

Mr. GREEN. Thank you, Mr. Chairman, for both you and our ranking member, Ms. Schakowsky, for having the hearing today. While the technology behind autonomous vehicles continues to evolve at a rapid pace it is important that industry and Congress continue to examine safety standards to ensure consumer safety. Not all the safety innovations are willingly accepted by the public with the history of airbags and seatbelts has shown. Continued open discussion on these new technologies are essential moving forward so that consumers can be familiar with both benefits and the limits of autonomous features. Frankly, my wife is probably the most supporter of me not being in an autonomous vehicle when I am driving. She complains all the time about my driving.

Mr. Zuby, in your testimony you state that your research has shown that the driver acceptance of technology varies. Can you tell us more about the varying level of acceptance of new technology and what can be done to increase the public's acceptance?

Mr. ZUBY. Yes. For one of the things that we found for lane departure warning systems, the mode of the warning made a big difference in whether or not the drivers accepted them. When we interview drivers what we find is they complain about audible warnings being annoying. Another important aspect of lane departure warning and lane maintenance is that the systems respond to truly dangerous situations and not be perceived by the driver as simply being a nanny about use of the turn signal.

So I think the technology needs to go a ways beyond where it is today in order to sort out what are the real dangerous situations that we need to inform the driver about versus those things that might be dangerous, but a lot of drivers aren't going to perceive them as such.

Mr. GREEN. OK. At this point, is it known why one warning system is so effective and another ineffective?

Mr. ZUBY. One of the issues is if the warning system can be heard by other people in the vehicle drivers tend not to like it. So the vibrating steering wheels, the vibrating seats tend to have higher levels of acceptance than audible warnings themselves.

Mr. GREEN. Thank you. How can we better study the effectiveness of these safety claims to ensure technology is living up to its promise?

Mr. ZUBY. It is super important I think that we work out ways to make sure that data about which cars have which systems and how the systems are working is available to independent researchers. Obviously, the companies who are developing the systems are going to want to make claims about their high levels of effectiveness, but I think people in Government and independent evaluators need to be able to verify those claims.

Mr. GREEN. I would like to ask this question of the entire panel. Would enhanced Government regulation on the collection of the crash data with specific regard to what autonomous technologies were in each vehicle improve both public safety and efficiency, the AV technology? I will start with Mr. Klei.

Mr. KLEI. Yes. Certainly when you look at things like the Auto ISAC, which has been developed as an industry coalition to really share data on cybersecurity, it is a good example where data sharing can really benefit. We think there is an opportunity as well to do something similar for some of the crash data and some of the activity around autonomous, automated driving vehicles. We think that the sharing is very powerful, but it needs to be the edge cases and it needs to be things that can help all of us develop and deploy these technologies.

Mr. GREEN. Mr. Gouse.

Mr. GOUSE. In our committees, sir, there is quite a bit of sharing going on of technical information that is not proprietary to build the standards to design test specifications, test devices, and what not to build good product, so there is a quite a bit ongoing already at that level.

Mr. GREEN. OK. Mr. Zuby.

Mr. ZUBY. Definitely, I think regulations prescribing what kind of data needs to be saved and under what kind of circumstances and with whom that data can be shared will help all of us achieve a greater level of comfort that the technology is being developed in a safe way.

Mr. GREEN. Dr. Stepper.

Dr. STEPPER. Definitely a yes, Congressman. Bosch has been working very adequately to actually get NHTSA more resources for data for crash reconstruction. Why, because we have used NHTSA's NASS database for our own research in understanding how many percent of collisions with injuries and fatalities with rear-end crashes, how many drivers failed to, for example, even after they received the warning to even apply the brakes in the first place. So it is very valuable data for us for our development purposes.

Mr. GREEN. OK. Thank you, Mr. Chairman. I yield back.

Mr. LATTA. Thank you. The gentleman yields back, and the Chair now recognizes the gentleman from Florida for 5 minutes.

Mr. BILIRAKIS. Thank you, Mr. Chairman. I appreciate it. Dr. Stepper, some driver assistance systems on the market use audible tones, steering wheel vibrations, and flashing lights to alert the drivers to impending hazards. We are also facing high levels of driver distraction as you know. As Bosch works to develop these technologies how are you working with automakers to ensure that these technologies aren't pulling drivers' attention away from the task of driving and causing more distraction?

Dr. STEPPER. Thank you, Congressman, for the question.

Mr. BILIRAKIS. Sure.

Dr. STEPPER. We work very intensively with our OEM partners on the human factors element. For example, evaluating what is a really effective and efficient means of alerting the driver of getting the attention from the driver back? Is it audible, is it visual, is it maybe haptic?

As Mr. Zuby has answered before what we have found is that haptic feedback is actually very, very efficient when it is related to a specific action that is wanted. For example, if there is a hazard approaching from the rear left, if your seat vibrates on the left side of the driver's seat there is a haptic feedback that alerts you that something is happening to the left of the vehicle. Or if it is intended that you are, for example, departing your road lane, the vibration of the steering wheel is directly related to something that is going on with the steering system that the driver should pay attention to.

We have formed our own group to work on human factors to specifically look at the human-machine in action and we work very intensively not only with our OEM customers but also with academia on this topic.

Mr. BILIRAKIS. Mr. Klei, do you want to comment on that as well?

Mr. KLEI. Yes, I think similar to the Bosch development we also have a very significant investment in the human-machine interface technologies. We have been one of the leaders in displays, in clusters, and in warning systems for vehicles for many, many years. We think that is an important part of bringing these technologies to market safely.

Clearly, when it comes to the audible versus haptic, we have done a lot of research as well. We actually have driver monitoring cameras that we are looking where the driver is seeing, or looking, where the driving task should be. And we sometimes use LED lights or other ways to try and bring the driver's attention back to the driving task. That is a big question.

As you talk about Level 3 technologies that is the biggest question and the biggest area of development is how do you get the driver disengaged and then re-engaged fast enough to resume the driving task. And I think that is a challenge for the industry. That is why you see some developing from Level 2 to Level 4, some are going to go through Level 3. But that is probably one of the biggest challenges and we are investing heavily in this area.

Mr. BILIRAKIS. OK. As a follow-up, are consumers able to manually turn off these alerts or warnings or customize them to their individual preferences?

Mr. KLEI. So that is really a question for the OEM to determine what they would like to do. And it happened as well with ABS and electronic stability control and the various traction control systems, the OEMs for many years could determine which could be turned on and off. So it is something that some allow, some don't. We believe that ultimately when it is proven that the safety technologies are really going to save lives that it shouldn't be turned off. It should be developed over time to be very easy to understand, very easy to use, and will ultimately save lives.

Mr. BILIRAKIS. OK. I have a question with regard to actually a follow-up on the gentleman from West Virginia. I mean, we want to help a lot of the elderly, maybe physically disabled people get around. We don't have in my area, in the Tampa Bay area we really don't have a mass transit system, so this could be extremely beneficial to people getting to doctors' appointments, what have you, these automated cars.

But you anticipate them having a standard driver's license; is that correct? I mean they have to qualify for this. For example, if you have a visual disability, if you are visually impaired and you don't qualify. I am visually impaired but I qualify at this particular time. I have a standard driver's license. I don't drive at night, but 5 years from now, who knows? Will I be able to drive one of these cars even though I am visually impaired? That is just an example there. Can I hear from one of you? What do you anticipate?

Mr. KLEI. Certainly we believe like we have talked a lot about the improvements in mobility for disabled and then certainly we think these technologies will offer significant improvements here. But it takes time and it takes really more, the systems that are developed with that in mind. And that is why we are working hard as a company with our OEM partners to make sure that these systems are developed with all considerations in mind. It is not just for the driver that has, you know, zero disabilities. It is to provide mobility for everyone. And we think there is a clear promise and they are being developed with this in mind.

Mr. BILIRAKIS. Anyone else?

Mr. GOUSE. May I, please. We have been working with AAMVA, the American Association of Motor Vehicles Administrators, on that exact topic for both cars and trucks. And a simple example would be some States require that parallel parking is required to get your initial driver's license, but in some vehicles the vehicle itself can parallel park without the assistance, with the assistance——

Mr. BILIRAKIS. If you could put the mike a little closer.

Mr. GOUSE. So we have been working with them trying to define what features are in place or are possibly in place in the future and they can design their driving tests and their ratings or perhaps certification levels like a commercial driving license has or something that says you can operate a Level 3 vehicle with these features, but you can't do a completely manual one. You can't drive a manual transmission anymore. So it is a complicated question, but it is being worked on.

Mr. BILIRAKIS. And there will be a State issue, obviously, as far as that is concerned. OK, well, that is important. I mean, we have got to know that ,because we want to help out our constituents. But again, you know, if you have a standard driver's license you qualify. And the gentleman asked about someone that is intellectually impaired. You know, would that person qualify? More than likely they couldn't get a license. So anyway that is something we have to resolve, so I appreciate that. I have one more question if I have time. I don't have time.

Mr. LATTA. Yes. If you would like to submit it in writing that would be great.

Mr. BILIRAKIS. Yes, I will submit it. Thank you very much. I yield back. Thank you.

Mr. LATTA. Thank you very much. The gentleman yields back, and the Chair now recognizes the gentlelady from New York for 5 minutes.

Ms. CLARKE. Thank you, Mr. Chairman, and I thank our ranking member. I thank our expert panelists for a very important and stimulating examination of autonomous cars.

Some experts have raised particular concerns regarding Level 3 automation and you have discussed it here today where a vehicle can drive itself but the driver must be ready to take over at a moment's notice. There is some evidence that Level 3 may lead to an increase in traffic collisions. During recent test drives, Ford reportedly noticed that even their engineers trained to monitor autonomous vehicles had trouble staying alert at the wheel while the car was driving. Volvo's autonomous vehicle program is skipping Level 3 altogether and planning to go straight from Level 2 to Level 4.

Mr. Zuby and Mr. Gouse, do you agree that complications of Level 3 automation are an example of why it is important to monitor autonomous technology to make sure that it is actually making driving safer?

Mr. ZUBY. Yes. Thank you for the question. Absolutely, I think the important thing will be to be able to monitor these developments as they are put out into the fleet. There is a long history of human factors research that says things like Level 3 are potential problems for human monitors, and I think that is why you find some automakers and some technology developers deciding that they aren't going to mess around with Level 3.

I am not expert enough to know that Level 3 is impossible to do successfully, but definitely there is a concern that if the car is too highly capable at the dynamic driving task that the driver will discontinue his monitoring activities and not be able to resume control when it is necessary because the system is no longer capable handling a situation.

Mr. GOUSE. I would just second what David said, but I would like to caveat with, bear in mind that people working on this—I am just awed when I go to committee meetings and listen in at the experts, the level of knowledge that is behind all this and the amount of consideration that is going on for all the aspects. Whether it be taking over control immediately or changes in weather conditions or road issues or anything at all these levels, it is very impressive the level of expertise and the care that is going into this.

Ms. CLARKE. The only factor that I guess is challenging to sort of pin down is human error, right?

Mr. GOUSE. Well, there are other challenges too, just like in our normal driving that we have unexpected issues that arise. The deer jumps out that you never saw before and how do you react to that? Or there is some sort of a failure in the vehicle or in the infrastructure that is unanticipated and how do you react to that? Or someone else who has not got automation or not got assistance and makes a grave error and how do you react to that?

Ms. CLARKE. But the reaction is the human being, right, not necessarily the vehicle? Or is it that the vehicle would be programmed to react to the jumping deer or the change in weather conditions?

Mr. GOUSE. Well, that goes back to the level of automation, whose job it is, who is it assigned and——

Ms. CLARKE. So Level 3 then becomes the challenge in terms of what the standard would be for automation versus human participation.

Mr. GOUSE. The expectations between Level 2 and 3, it is a big step.

Ms. CLARKE. OK. As we have heard, semi-autonomous features can have significant safety benefits but they may also be confusing, especially to drivers who are unfamiliar with the technology or fail to use it correctly. Consumer education will be essential to ensuring that the full advantages of these technologies are realized.

Mr. Zuby, why is it so important that drivers understand that limits of semi-autonomous features and are aware of what exactly their cars can and cannot do?

Mr. ZUBY. Yes, for exactly the issues that we have been discussing about Level 3. I mean it will be important for drivers to understand how close attention they need to pay to the driving situation in order to be ready to take over and wonder what situations the system is likely to hand control back to them.

But we would say that I think it is important to try to figure out how to design these things so that the limitations and the way they work is as intuitive as possible because I don't think we can rely on people to spend extra time to learn how to drive their cars. I mean how many people in this room have read their owner's manual from front to start? There is a lot of really important information in there, but I for one have not read the owner's manual from start to finish for any of the vehicles I have ever owned.

Ms. CLARKE. Very well. Mr. Chairman, I yield back.

Mr. COSTELLO [presiding]. Mrs. Walters.

Mrs. WALTERS. Thank you, Mr. Chair.

Mr. Gouse, we know that many States and localities have developed legislation aimed at regulating self-driving cars. Can you go into further detail on the State localities implementing SAE's level of driving automation into their laws?

Mr. GOUSE. I am most familiar as a staff person with Pennsylvania and Michigan and California. But there are, as Jeff said earlier, there are two or three dozen States, and at each State or Commonwealth there is an upper chamber and a lower chamber and also there may be a regulatory agency, or two of them that are working in concert or in parallel paths. So there are quite a few going on.

And our members who are active are picking up things. I know New Jersey is talking about it. I heard that from a member yesterday. North Dakota is a State, I believe. So it is not our main business as SAE to monitor State activities, but we want them to adopt the SAE language so there is consistency across all the States and territories.

Mrs. WALTERS. Yes. I think that is going to be an issue. The consistency is going to be obviously very, very important. And then the same question for you again is a number of groups have developed classification systems to define automated driving systems, and can you discuss why SAE determined the J3016 standard to be the most optimal way of defining the different automated driving systems?

Mr. GOUSE. I would just like to say probably that the committee leadership and members worked very hard on this over quite a bit of time with a tremendous amount of input from various different stakeholders. And it is not just a committee of technology developers, there are policy folks in there, NHTSA was part of it, motor carriers, Federal Motor Carriers was part of it.

So it was an ongoing process. It was in fact adopted internationally before NHTSA did even at the Amsterdam convention in April of '16, I believe. So it is becoming a global standard and it is being validated that way across the globe and in the States as being the preferred choice. It is also a living document. It has been revised already once since it was issued. In fact, the name was even changed a little bit to clarify it. So it will go through revisions and additional references to discuss some of the issues that were brought up here in questions to add to it.

Mrs. WALTERS. OK, all right. Thank you very much, and I yield back the balance of my time.

Mr. COSTELLO. Mr. Cárdenas, you are now recognized for 5 minutes.

Mr. CÁRDENAS. Thank you very much, Mr. Chairman. Something just occurred to me. Are we likely going to see in the near future—I grew up learning how to drive on a stick shift. A lot of today most drivers in America probably don't know how to use a manual or a stick shift vehicle, these automatic gear shifting vehicles. Are we looking at possibly in the near future where people get in their car and they push a button, today I am going to use automation 1, 2, or 3 Level, and maybe that is the new gear shifting or shifting of the vehicle that we are going to be driving in the future? Does that make any sense, or is that probably likely what we are going to be looking at?

Mr. KLEI. I think one of the things that we look at when we are looking into development is you never take the fun away from driving your car. We still like the ability for people to drive their cars when they want to drive their cars. But there is many driving tasks, there is many opportunities for disabled to provide mobility, and that is where we think the big benefit will be. We never want to take the fun away though.

So it could be someone gets in a car and says yes, I want to go from point A to point B in an automated way or it could be that I want to drive myself on the windy country roads. So I think there is going to be some opportunities there over time for people to still have fun, but in certain circumstances still get the mobility that they need and they want and to be able to do other things in the car.

Mr. CÁRDENAS. Well, speaking of taking the fun away driving, I can envision if we are going to be appropriate as a Government, and maybe in the future what we have is a speed limit technology where if you are going to be driving an automated vehicle then the speed limit is 35 miles an hour. Your car is not going to be allowed to go over 35 miles an hour on that piece of the road.

Mr. KLEI. Yes. I mean, I think these are things that we need to consider, but quite frankly we believe that if you do that you could actually introduce more challenges because everyone will try and go around the car. You want the car to flow naturally with traffic with other automated vehicles as well as nonautomated vehicles, so you want it to be very natural, and through testing and development that is what we are developing for. So to limit a car and limit the mobility and limit the functionality is going to limit the testing and deployment of such technologies and potentially lifesaving benefits.

Mr. CÁRDENAS. For those of you who are on the panel from private industry, I mean how do you feel about your relationship right now with Federal departments when it comes to reporting and expectations of, you know, obviously nonproprietary progress and letting them know what you are looking for as long as timing of introducing products, et cetera?

Mr. KLEI. I think, Congressman, it is a great question. It is one that through the Federal Automated Vehicles Policy that was rolled out last September from NHTSA it is a great start to bringing the collaboration together between industry and Government. And we think it is a big step forward but there is more work to do.

In that policy it requires significant reporting between the industry and NHTSA and that reporting needs to be better defined, it needs to be more expedited, and the exemption rules that we are all looking for especially in the development side need to be improved. And so we are working closely with that agency, with NHTSA to try and improve that and make sure that when it is officially rolled out and deployed that really it is, in fact, usable and it is going to drive this technology forward and potentially save lives when deployed.

Mr. CÁRDENAS. What country right now seems to be more, I don't want to use the word advanced, but more ready and willing to allow their constituents to drive the highest class of automated vehicle right now?

Mr. KLEI. Every country has certain limitations and certain regulations and there is no one country that is easy. Every country has different——

Mr. CÁRDENAS. I mean, is there a particular country right now that—I am thinking of Germany. I am wondering if they are allowing a little bit more than we are so far.

Mr. KLEI. I don't know that there is one country that says it is easy to do. Every country has certain limitations and for good reason.

Mr. CÁRDENAS. Anybody know what is going on around the world?

Dr. STEPPER. The same as Mr. Klei said, from my side sometimes it is not even regulated by a specific country law. You know, also in Germany, you mentioned Germany as an example, the different States have different laws and different regulations and the regards of allowing or not allowing different levels of automation. There may be some States that are really fostering the rollout so that companies like Bosch can go on public roads and test and validate the systems which is very helpful for our development to be allowed to do that.

Mr. CÁRDENAS. OK. Thank you, Mr. Chairman.

Mr. COSTELLO. Mr. Mullin, you are now recognized for 5 minutes.

Mr. MULLIN. Thank you, Mr. Chairman.

Doctor, is it Stepper?

Dr. STEPPER. Yes.

Mr. MULLIN. Thank you for being here. You talk about the technology and moving forward with the technology of going out and testing the vehicles. But can you explain a little bit more how that works with the technology of the vehicle versus the GPS——

Dr. STEPPER. Yes.

Mr. MULLIN [continuing]. That the vehicle I am assuming has to be programmed into a GPS and it has got to take you from point A to point B; is that correct?

Dr. STEPPER. So it depends on the level of automation, Congressman. So if you would go all the way to a Level 5 automated driving, for example, which really takes the driver out of the loop and there is no longer a driver required to operate the machine that it would exactly the scenario that you would dial in a particular destination and the vehicle will take you there, for example, door to door.

Mr. MULLIN. Well, what is Level 1?

Dr. STEPPER. In Level 1, this is what we call today's driver assistance systems where there is——

Mr. MULLIN. Where your seat vibrates and it tells you and does all that stuff?

Dr. STEPPER. For example, there would be a warning that there is an impending front-to-rear-end collision or there is a lane departure that is about to happen.

Mr. MULLIN. And 2?

Dr. STEPPER. Two combines the longitudinal and lateral control of the vehicle so, for example, we still call it the assistance functions. It is functions like a traffic jam assist where the vehicle in that particular scenario in a traffic jam would automatically take the control for the longitudinal and the lateral perspective of the vehicles but the driver is still fully responsible and fully in the loop, whereas in Level 3, for example, you take that as one example to a traffic jam pilot where you can take your hands and your feet off for a well-defined scenario.

You need to be on a Class 1 road. On a traffic jam pilot, for example, you need to have preceding traffic, and then for this stop and go traffic the machine would take over the control of the vehicle until it handles it back to the human being.

Mr. MULLIN. And 5 is what we started the conversation with. Do we see the advancement of the vehicles catching up or going to surpass the GPS? Because everybody uses their road maps and their GPSs on their phones and I am sure I am not the only one that it takes me to the wrong place all the time.

Dr. STEPPER. Yes, yes.

Mr. MULLIN. So they would have to work simultaneously, wouldn't they?

Dr. STEPPER. Yes, so they actually, Congressman, there is additional technology that is required. So what we know today as GPS, also standard definition maps, for Level 4, Level 5 automated driving to a certain extent even for Level 3, we have the need for high resolution, highly dynamic maps that really exceed the requirements that we see from the map requirements from today's navigation system. And that is actually coupled in a process called data fusion with onboard sensing via radio cameras, your radars, your other sensing technology you may have on board on the vehicle that will recognize certain landmarks like a fire hydrant, like a bridge, like a certain exit, and it combines the GPS information——

Mr. MULLIN. That is more of an eyesight on it.

Dr. STEPPER. As well as nonvisible electromagnetic base like radar, for example, or LiDAR technology which uses laser light.

Mr. MULLIN. So would this be one entity or would each company be responsible for their own technology for the GPS to which their vehicle is going to be operating by?

Dr. STEPPER. It really comes together at the end at the vehicle manufacturer. There may be different suppliers for certain sensing technologies or GPS technology. What really is the trick to have the competency in bringing all this data together in this data fusion process and derive driving policy decisions out of that.

Mr. MULLIN. What I am talking about is somebody working on this end of the GPS as you guys are working up with the vehicle, are they going to meet? Or when the technology for the vehicle gets to that point, then we start diving into the precise GPS?

Dr. STEPPER. Yes, so that is already available today in a system that is called differential GPS systems that increases the resolution. Most companies, actually, out there testing and validating automated driving today use differential GPS system to get them to the resolution that they need, which in essence is a centimeter resolution as opposed to a couple meters that we see today. So that technology is already available today. The challenge in the development is going to be to bring the prices down and the costs down of such an advanced GPS system for use in every vehicle.

Mr. MULLIN. Is there one company that is leading that?

Dr. STEPPER. There are several companies that are working on that exact topic. There is not one company that stands out.

Mr. MULLIN. Do you have one particular one that you are working with?

Dr. STEPPER. We work really with all of them at the moment. There is no particular one that I can point out at the moment, Congressman.

Mr. MULLIN. All right, thank you. Thank you for your time. Mr. Chairman, I yield back.

Mr. COSTELLO. Thank you. I will now recognize myself for 5 minutes and ask a question to all panelists, two-part question: One, how is the development and testing of these systems different from the development and testing of fully self-driving technologies; and second, how much can be learned from the development and testing of advanced driver assistance systems?

Mr. KLEI. So first, what is different, I don't really think there is so much difference in the way we develop and we test technologies, everything from ABS through electronic stability control and all the way to fully automated driving. It is a very rigorous, long testing process. It starts with the technology itself. It starts with bench testing, then in contained track environments, and we evolve all the way to, ultimately, the real road and real world testing.

So the process is very similar. Obviously, the conditions by which we test are going to be different depending on the technology. But in terms of the rigorous, you know, Six Sigma, continuous improvement mindset that we have to make sure the products are safe is no different regardless of what the technology is. The challenges are bigger the higher levels of automation you go to, but the testing process itself is always very much the same, safety first.

When it comes to the implementation of these and across the various product portfolio again everyone is going to be different, and ultimately it is the OEM that decides when it is safe to deploy in the vehicle. We work with OEM customers and they ultimately are the ones that certify for FMVSS.

Mr. GOUSE. I would like to just briefly add a couple things. Prior to the beginning of testing, there are some tools you put in place, what are called a design failure mode effects analysis and failure mode effects analysis, where you look at all different ways a system might fail and then you design a test procedure to encompass that and then you look at when something fails, whether it is part of the system or something external or you are testing an automated vehicle, but the engine conks out or something or you get a flat tire, you have to build all of that into your test procedures. And so you have got a complete, very comprehensive, and carefully designed program to execute as part of the process.

Mr. ZUBY. Yes. I would agree with Mr. Klei and Mr. Gouse that the process is similar. But I think one of the things that we need to keep in mind that as we deploy increasingly evolving technologies we do need to watch them very carefully and see how they perform in the real world. And when they fail to perform try to understand whether or not they are failing to perform because of a deficiency in the technology, a deficiency in the logic behind the technology, or because the circumstance in which they failed is just outside the design domain of that particular technology.

Again, consequently, I think information about what is happening in the real world as these technologies deploy is going to be vitally important to making sure that this stuff is developed in a safe way.

Dr. STEPPER. And if I just may add a few points. Number one is what we didn't have available in the past when we started developing ABS or ESP, for example stability control, was an international standard specifically designed for the different safety assessments and different safety levels. And that standard is called ISO 26262 which was specifically developed for use in the automotive space to define different safety levels and also define how to get to and what you have to meet in order to get to the different levels of this safety.

Number two, what we didn't have available when we are deploying ABS or electronic stability control or early in driver assistance is the vehicle being connected to the rest of the world, being connected to servers. If we would just proceed with conventional validation as we have in all these decades it would really be cost and time prohibitive. We would; literally, in order to fully validate a fully automated vehicle we would have to drive a distance that equals the average distance between the sun and the earth which is not feasible from a cost and time perspective.

So what we continue to deploy is the advantages of being connected and having vehicles deployed in the field that collect for us very valuable data of real world traffic situations that we then can take back to analyze and develop and adjust our software, for example, accordingly.

Mr. COSTELLO. Thank you. Seeing there are no further members seeking to ask questions for the first panel, I would like to thank all of our witnesses again for being here today.

Before we conclude, I would like to include the following documents to be submitted for the record by unanimous consent: a report from MEMA; Advocates for Highway and Auto Safety's FAVP comments in a March 27th letter to Chairman Latta and Ms. Schakowsky; a statement from the National Safety Council; a statement from Global Automakers; a letter from the U.S. Chamber of Commerce, Technology; a statement from American Car Rental Association; a statement from Mobileye; a statement from EPIC; and a letter from Honda.

[The information appears at the conclusion of the hearing and at *http://docs.house.gov/Committee/Calendar/ByEvent.aspx?EventID=105790.*]

Mr. COSTELLO. In pursuant to committee rules, I remind members they have 10 business days to submit additional questions for the record and I ask that witnesses submit their response within 10 business days upon receipt of the questions. Without objection, the subcommittee is adjourned.

[Whereupon, at 12:00 p.m., the subcommittee was adjourned.]

[Material submitted for inclusion in the record follows:]

ADVOCATES
FOR HIGHWAY
& AUTO SAFETY

March 27, 2017

The Honorable Robert Latta
Chairman
Committee on Energy and Commerce
Subcommittee on Digital Commerce and
Consumer Protection
2125 Rayburn House Office Building
Washington, D.C. 20515

The Honorable Jan Schakowsky
Ranking Member
Committee on Energy and Commerce
Subcommittee on Digital Commerce and
Consumer Protection
2125 Rayburn House Office Building
Washington, D.C. 20515

Dear Chairman Latta and Ranking Member Schakowsky:

As you prepare for tomorrow's hearing, "Self-Driving Cars: Levels of Automation," Advocates for Highway and Auto Safety (Advocates) would like to submit our position on the safety implications presented by autonomous vehicles (AVs). Advocates is a coalition of public health, safety, and consumer organizations, insurers and insurance agents that promotes highway and auto safety through the adoption of safety laws, policies and regulations. We respectfully request that this letter and the comments Advocates submitted to the public docket in response to the National Highway Traffic Safety Administration (NHTSA) "Federal Automated Vehicles Policy" (AV Guidelines) Notice and Request for Comments (81 Federal Register 65703, September 23, 2016, DOT Docket No. NHTSA-2016-0090), which are attached, be included in the hearing record.

Advocates Has Consistently Pushed for Advanced Technologies in Vehicles to Save Lives and Prevent Injuries. With Fatalities on the Rise, Action is Needed.

Advocates has been a long-standing leading supporter of technological solutions to advance safety, reduce crashes, save lives, mitigate injuries and contain crash costs. These efforts include promoting requirements for airbags, electronic stability control, anti-lock brakes, rearview cameras and other important safety features as standard equipment on cars, trucks and motorcoaches. In fact, NHTSA has estimated that since 1960, over 600,000 lives have been saved by motor vehicle safety technologies.[1] Autonomous vehicle technology presents a similar possibility of accomplishing significant reductions in preventable motor vehicle deaths and injuries at a time when fatalities are on the rise.

According to NHTSA, 2015 experienced the largest percentage increase of motor vehicle deaths in nearly fifty years.[2] More than 35,000 people were killed on our nation's roads, representing a 7.2-percent upturn.[3] Preliminary information for the first half of 2016 appears to be even worse, indicating an 8 percent rise in fatalities compared to the same time period in 2015.[4] Advocates is optimistic that AV technologies can help reverse this recent trend.

---

[1] Lives Saved by Vehicle Safety Technologies and Associated Federal Motor Vehicle Safety Standards, 1960 to 2012, DOT HS 812 069 (NHTSA, 2015); See also, NHTSA AV Policy, *Executive Summary*, p. 5 endnote 1.

[2] National Center for Statistics and Analysis, *2015 motor vehicle crashes: Overview*, Report No. DOT HS 812 318, National Highway Traffic Safety Administration (Aug. 2016).

[3] *Id.*

[4] National Center for Statistics and Analysis, *Early Estimate of Motor Vehicle Traffic Fatalities for the First Half (Jan–Jun) of 2016*, Report No. DOT HS 812 332 (Oct. 2016).

Yet, some experts forecast that 15-20 years may transpire before AVs comprise a major portion of the vehicles on public roads. In the interim it is unacceptable to complacently allow more than 500,000 people to be killed and more than 36 million to be injured in crashes. In the short term, we urge NHTSA to use its authority to require that available and effective crash avoidance technologies be required as standard equipment on all motor vehicles. These include automatic emergency braking (AEB) and lane departure warning systems for trucks, buses and cars. To encourage these advances, Advocates, along with other safety groups and families of victims and survivors of crashes, filed a Petition for Rulemaking with NHTSA in 2015 requesting the agency issue a rule to require automatic braking systems to prevent frontal crashes involving large trucks. The agency granted the petition and Advocates urges NHTSA to commence rulemaking this year, particularly because of the urgency in addressing the unacceptable and dramatic increases in truck crash deaths these past five years.

Semi-Autonomous Vehicles that Share Control with the Human Driver Pose Serious Safety Challenges.

Regarding the specific focus of the Subcommittee's hearing, the levels of autonomous operation of AVs, the Society of Automotive Engineers (SAE) and NHTSA in its AV policy[5] have adopted a range of levels, 1 through 5. Levels 1 through 3 involve some form of safety-oriented technology or AV system that may only alert the driver by providing a warning, assist the driver in taking evasive action, control a particular safety system in order to prevent a crash, or operate the vehicle in certain circumstances while the driver is supposed to monitor the vehicle. These three levels represent varying degrees of reliance on independent automated technology systems but can also involve shared control of the vehicle by the driver and the AV operating system at different points in a trip. At each level of autonomous or semi-autonomous operation through level 3 the driver must remain completely engaged in the driving task. The driver must remain alert, monitor the vehicle operation and driving task, and either maintain control of the operation of the vehicle or be prepared to take control of vehicle operation (re-engage) in the event the AV system fails to function properly or cannot respond (shuts off) under the circumstances prevailing at the time. This shared control poses serious safety challenges as drivers may become overconfident and allow themselves to be distracted and/or lulled into a false sense of security by the AV system.

An example of the consequences that can occur when an AV system is not properly vetted and tested was the May 2016 crash in Florida of a Tesla Model S using the Autopilot AV system that resulted in the death of the vehicle owner. First, while the Autopilot system was designed to have the driver constantly monitor the operation of the vehicle, the AV system method for maintaining driver engagement was insufficient. Reminders to the driver to keep his hands on the steering wheel were inadequate and too far apart in time sequence to ensure driver re-engagement during a critical safety event. Second, the vehicle visual sensors for the AEB system did not identify that a large truck had crossed the path of the Tesla and presented an immediate danger. The radar may have detected the truck but dismissed it. The conflicting inputs from the camera and radar sensors did not trigger any safety action by the Autopilot AV system, such as switching off the vehicle cruise control or applying the AEB system. Despite the conflict in sensor information, and the lack of response by the driver to any driver engagement warnings, the autopilot remained engaged and drove the vehicle under the truck killing the driver. This example of a shared responsibility for vehicle operation by the driver and the AV system, and the hand-off that needs to occur between the AV system and the driver, clearly show that Level 2/3 AV systems present a particularly high degree of safety risk to the public. This is the reason adequate testing of the AV operating system is essential in order to ensure that drivers

---

[5] The NHTSA AV policy identifies levels 3-5 as highly autonomous vehicle (HAV) operation.

remain engaged in the driving task and that the transition from autonomous operation to driver re-engagement works perfectly every time.

Fully Autonomous Vehicles Present Unique Safety Concerns.

AV operation levels 4 and 5, which represent fully autonomous operation controlled by the AV system, present a different set of safety concerns. Level 4 and 5 vehicles must be able to complete a trip entirely based on AV system control, without driver input. To do so, the AV system must undergo rigorous and thorough testing to ensure it is capable of operating flawlessly. In the event of a mechanical or software problem, defect or failure, the AV system must be able to put the vehicle in a safe mode that takes it out of harm's way.

A Functional Safety Approach is Needed to Provide the Framework for the Design, Development, and Deployment of Autonomous Vehicle Technology.

Advocates' comments to the docket on the AV Guidelines urged NHTSA to require a functional safety process for new AV technologies which are rapidly entering the marketplace. The command and control software of these vehicles is not addressed by current Federal Motor Vehicle Safety Standards (FMVSS). Furthermore, it is expected that there will be efforts by industry to seek and obtain exemptions for AVs from some or all of the existing FMVSS. While we know there will be crashes, deaths and injuries during the transition between old and new cars, human error should not be replaced with computer error. Predictable problems and flaws that pose unreasonable risks to public safety before these vehicles are sold to the public and used on public roads should be eliminated, and this can be achieved with a mandatory functional safety process.

Additionally, cybersecurity is an important aspect of AV development which must be addressed as part of functional safety. NHTSA should identify problem areas and require specific responses from manufacturers as to how those are being addressed. Problem areas could include subjects such as GPS signal loss or degradation, spoofing, and off-line and real time hacking of single vehicles or fleets of vehicles. As with all other AV performance aspects, the sharing of data in terms of cybersecurity will improve overall safety and ensure that all vehicles are afforded the same level of security. Data and information about known flaws or problems must be shared among manufacturers and with NHTSA and the public to ensure solutions to safety problems are readily identified and remedied. The potential risk of a single software error, or malevolent computer hack impacting hundreds or thousands of AVs, perhaps whole model runs, makes strong cybersecurity protections a crucial and essential element of AV design.

The Development of Autonomous Vehicles Must Be Transparent or Public Confidence in the Technology Will Suffer.

The development and deployment of AVs as well as NHTSA's role in regulating this technology must be open and transparent. All communications and responses between the agency and a manufacturer as it relates to any issues involving AVs must be made available for public review and scholarly research. All data generated from the testing and deployment of AVs, except for trade secrets and private individual information must also be made public. Lack of transparency will severely undermine the public's confidence in this new technology and inhibit its widespread adoption.

In fact, a recent national survey commissioned by Kelley Blue Book found that a large portion of the public is hesitant to accept AVs. Fifty-one percent of respondents replied that they prefer to have full control of their vehicle, even if it's not as safe for other drivers. Additionally, awareness of the higher levels of vehicle autonomy is limited, with 6 out of 10 people saying they know little or nothing about AVs. For half of the respondents, the perception of safety and personal comfort with autonomous technology diminished as the level of autonomy increased. In fact 80 percent believed that people should always have the option to drive themselves, and nearly one in three respondents said they would never buy a level 5 vehicle.[6]

Advocates urges the agency to issue appropriate safety standards for the AV technologies and we hope that the industries involved will support this position. Having some basic rules of the road that everyone follows, drivers as well as manufacturers, will benefit the auto and tech industries as well as public safety. If a lack of transparency or malfunctioning technology leading to crashes, deaths and injuries disrupts consumer confidence, it will set back all of our efforts to advance these lifesaving technologies.

Conclusion

In response to concerns about the death and injury toll on our highways, Congress passed the National Traffic and Motor Vehicle Safety Act of 1966.[7] The law required the federal government to establish the FMVSS to protect the public against "unreasonable risk of accidents occurring as a result of the design, construction or performance of motor vehicles."[8] While cars have changed dramatically over the last half century and will continue to do so in the future, the underlying premise of this prescient law has not. Technological advances, including AVs, offer the promise of achieving much-needed safety improvements. However, it is critically important that safety and transparency are at the forefront of the process.

Thank you for your time and consideration of our safety position on this emerging issue. Please do not hesitate to contact us if we can provide any additional assistance to the Subcommittee.

Sincerely,

Jacqueline S. Gillan
President

Catherine Chase
Vice President of Governmental Affairs

cc: Members of the Subcommittee on Digital Commerce and Consumer Protection

---

[6] 2016 Kelley Blue Book Future Autonomous Vehicle Driver Study, www.kbb.com.
[7] Pub. L. 89-563 (Sept. 9, 1966).
[8] Title 49, U.S.C. Sec. 30102.

[Additional information from AHAS is available at *http://docs.house.gov/meetings/IF/IF17/20170328/105790/HHRG-115-IF17-20170328-SD004.pdf*.]

Statement of the National Safety Council
House of Representatives
Committee on Energy & Commerce
Digital Commerce and Consumer Protection
Hearing on
Self-Driving Cars: Levels of Automation
Tuesday, March 28, 2017

Chairman Latta, Ranking Member Schakowsky and members of the subcommittee, thank you for allowing the National Safety Council (NSC) to submit this statement for the record. NSC is a 100-year-old nonprofit based in Itasca, IL, with a vision to end preventable deaths in our lifetime at work, in homes and communities and on the road through leadership, research, education and advocacy. Our more than 13,500 member companies represent employees at more than 50,000 U.S. worksites. For decades we have advocated for safer cars, safer drivers and a more forgiving environment in and around vehicles. We have led large scale public education campaigns on the importance of seatbelts and airbags, eliminating distracted driving, and helping consumers understand the technologies in their vehicles to reduce deaths and injuries on our roadways.

Federal leadership on motor vehicle safety is necessary because there should only be one level of safety. Consumers need confidence in vehicles regardless of where they reside; manufacturers need certainty in order to invest in design and production, and states do not possess the expertise and the resources to replicate design, testing and reporting programs. Further, a patchwork of requirements will result in confusion for consumers and increased cost for manufacturers and operators attempting to comply with a myriad of requirements. Finally, the absence of a safe, workable standard will drive development, testing and deployment overseas, resulting in the flight of innovation and the jobs that accompany it to locations outside of the US.

The Lifesaving Potential of Advanced Technology

NSC believes advanced vehicle technology, up to and including fully automated vehicles, can provide many benefits to society. The most important contribution will be the potential to greatly reduce the number of fatal crashes on our roadways, which are increasing. Every day we lose more than 100 people in motor vehicles crashes, and every year more than 4 million people are injured. Beyond the human toll, these deaths and injuries cost society over $380 billion, including productivity losses, medical expenses, motor vehicle property damages and employer costs.[1]

NSC preliminary estimates reveal that the 40,200 roadway fatalities during 2016 are 6% higher than the same period last year and 14% higher than the same period two years ago. If we are to

---

[1] Injury Facts 2017

make a meaningful change in this trend, there must be a sense of urgency coupled with large, near term gains to save lives on our roadways.

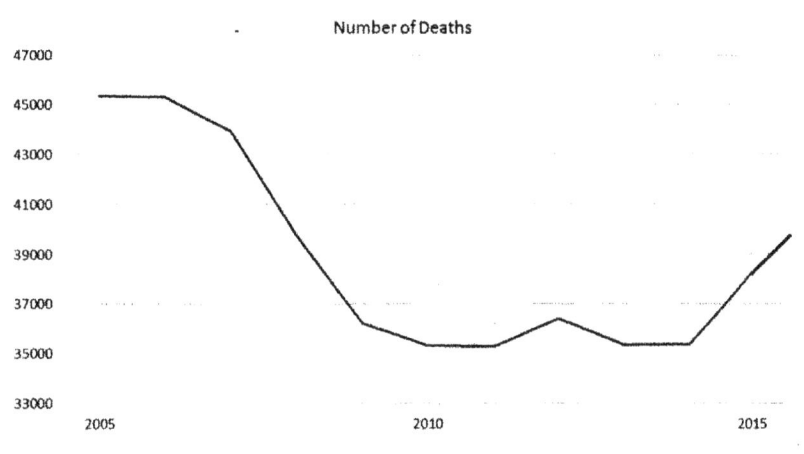

Motor Vehicle Deaths On the Rise

Source: NSC analysis of National Center for Health Statistics (NCHS) mortality data and NSC estimate for 2016

While the absolute numbers of fatalities change from year to year, many of the same behavioral problems remain persistent and have been represented in the data for decades. For example, in 2015:

- 9,306 people were killed in alcohol-impaired driving crashes[2]
- 3,477 people were killed in distraction related crashes[3]
- 9,874 people were killed while unrestrained.[4]

The National Highway Traffic Safety Administration (NHTSA) estimates that 94% of all fatal crashes have an element of human error. Therefore, if we are to eliminate or reduce the number of fatalities on our roadways, advances in vehicle technology must be part of the solution. However, it will likely be decades before we have meaningful fleet penetration of fully automated vehicles.

Last year, the NSC and the National Transportation Safety Board (NTSB) hosted a full day event with dozens of expert panelists focused on Reaching Zero Crashes: A dialogue on the Role of Advanced Driver Assistance Systems (ADAS).[5] While there is a great deal of excitement about highly automated vehicles (HAVs), automated vehicles and their potential to save lives, it is important to recognize that many legacy technologies represent the building

---

[2] https://crashstats.nhtsa.dot.gov/Api/Public/ViewPublication/812348
[3] NSC analysis of NHTSA FARS data
[4] https://crashstats.nhtsa.dot.gov/Api/Public/ViewPublication/812374
[5] http://www.ntsb.gov/news/events/Pages/2016_dta_RT_agenda.aspx

blocks for fully automated vehicles. Greater consumer acceptance of the dozens of safety technologies that are available today would lead to more rapid adoption of them, saving lives and preventing injuries.

As an example, Electronic Stability Control (ESC) is a technology that uses automatic computer controlled braking of individual wheels to help the driver maintain control in risky driving scenarios. ESC primarily mitigates single vehicle, loss of control crashes in which drivers would run off the road. For passenger cars as well as light trucks and vans, it is estimated that ESC systems have saved more than 4,100 lives during the 5-year period from 2010 to 2014, but incorporation into vehicles on the road remains slow.[6] The following charts from the Highway Data Loss Institute (HDLI) reveal how slowly ADAS technologies are achieving penetration in the US fleet due to normal turnover of inventory—with the average age of cars in the US fleet being 11.5 years old.[7] Electronic stability control has been available for decades and was mandated on all new passenger cars by the 2012 model year, but in 2015 only 40% of registered vehicles were equipped with ESC. Despite a clear life-saving benefit, full fleet penetration of this technology is not predicted until the 2040s.[8]

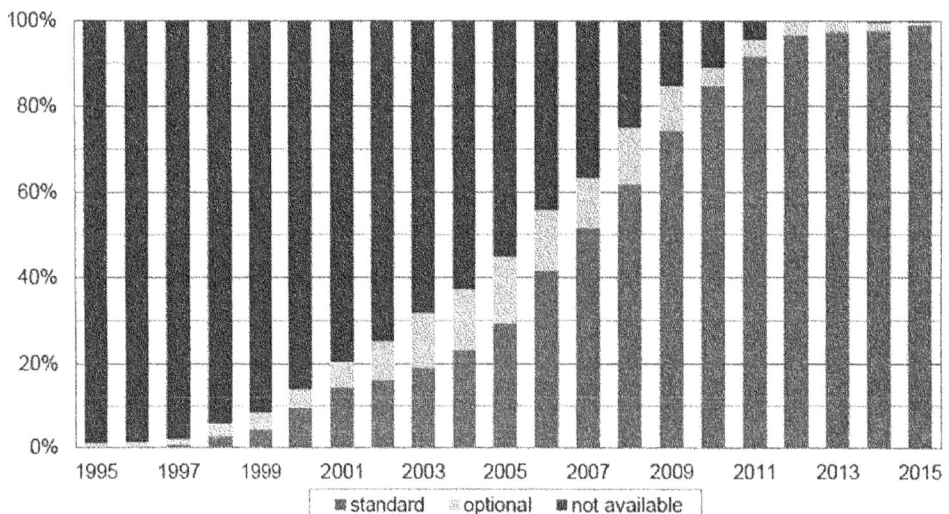

New vehicle series with electronic stability control
By model year

Source: HLDI

---

[6] https://crashstats.nhtsa.dot.gov/Api/Public/ViewPublication/812277
[7] http://www.rita.dot.gov/bts/sites/rita.dot.gov.bts/files/publications/national_transportation_statistics/html/table_01_26.html_mfd
[8] http://www.ntsb.gov/news/events/Documents/2016_dte_RT_p1_p3_moore.pdf

Registered vehicles with electronic stability control
By calendar year

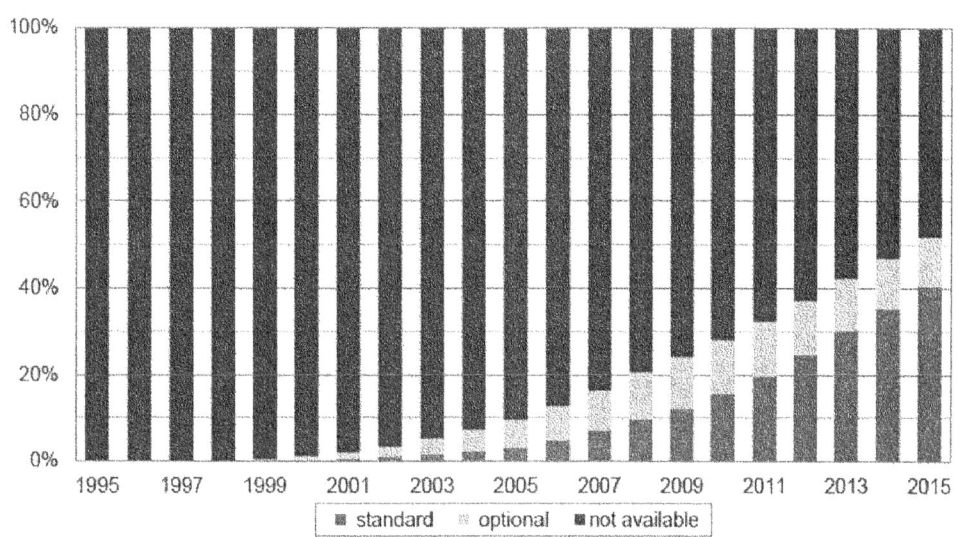

Source: HLDI

Registered vehicles with available ESC,
actual and predicted

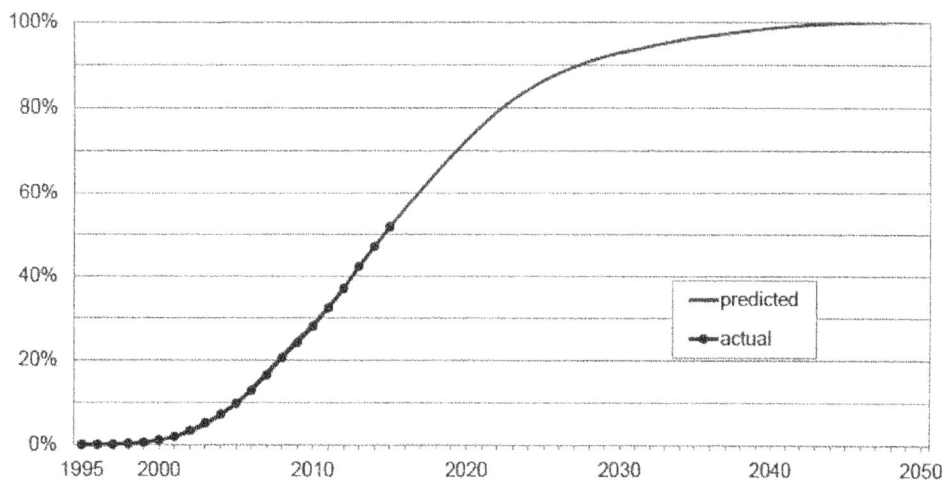

Source: HLDI

ADAS already operate on the roadways today, but more could be done to encourage greater fleet penetration. Features like lane departure warning systems, blind spot monitoring, adaptive cruise control and others help to prevent or mitigate crashes. The cost of these technologies is declining and their impact is measurable. According to the Insurance Institute for Highway Safety (IIHS), if four current technologies—forward collision warning/mitigation, lane departure warning/prevention, side view assist/blind spot monitoring, and adaptive headlights—were deployed in all passenger vehicles, they could prevent or mitigate as many as 1.86 million crashes and save more than 10,000 lives per year.[9] However, front crash prevention, commonly referred to as automatic emergency braking, which was an option in about half new 2015 model year cars, was in only 8% of registered cars in 2015.[10]

Crashes relevant to 4 crash avoidance systems
FARS and GES, 2004-2008

	all	injury	fatal
front crash prevention	1,165,000	66,000	879
lane departure prevention	179,000	37,000	7,529
side view assist	395,000	20,000	393
adaptive headlights	142,000	29,000	2,484
total unique crashes	1,866,000	149,000	10,238

Source: Insurance Institute for Highway Safety

Similar conclusions were reached in a July 2016, Carnegie Mellon study which stated that just three technologies—forward collision warning, lane departure warning and blind spot monitoring—could have prevented or reduced as many as 1.3 million crashes annually and over 10,000 fatal crashes.[11] This study further found that almost one quarter of all crashes could be affected by these crash avoidance systems, but only 2% of 2013 model year cars included these systems as standard.

While many of these technologies are available on higher value cars or as part of an upgraded technology package today, they are not standard equipment on all makes and models. Safety should not be just for those who can afford it, especially for technologies that will result in thousands of lives saved every year. The Carnegie Mellon study estimated that if all light-duty vehicles were equipped with the three technologies, they would provide a lower bound annual benefit of about $18 billion. With 2015 pricing, it would cost about $13 billion to equip all light-duty vehicles with the three technologies, resulting in an annual net benefit of about $4 billion or a $20 per vehicle net benefit. By assuming all relevant crashes are avoided, the total upper

---

[9] http://dx.doi.org/10.1016/j.aap.2010.10.020
[10] http://www.ntsb.gov/news/events/Documents/2016_dtc_RT_p1_p3_moore.pdf
[11] http://dx.doi.org/10.1016/j.aap.2016.06.017

bound annual net benefit from all three technologies combined is about $202 billion or an $861 per vehicle net benefit, at current technology costs.

NSC recognizes and applauds the voluntary commitment made last March by 20 automakers to include automatic emergency braking (AEB) on all vehicles sold in the US by 2022. Toyota has already committed to beat this date by several years. Given the slow turnover of the fleet, we encourage other manufacturers to view the 2022 date as a finish line rather than a starting point and accelerate the roll out of AEB and other lifesaving technologies.

Whether mandated or optional, in many cases these systems can perform driving tasks more predictably, more conservatively and more safely than a human driver, and may act without driver input if a driver is distracted, impaired or incapacitated. However, because there are no minimum standards for many of these technologies, legitimate questions about their effectiveness remain.

Dedicated Short Range Communication (DSRC)

Another component of ADAS and automated vehicle systems is dedicated short range communication (DSRC), which would allow vehicles to communicate over dedicated spectrum bands with each other, pedestrians, and infrastructure to prevent collisions. This technology, often referred to as V2V (vehicle-to-vehicle), V2I (vehicle-to-infrastructure), V2P (vehicle-to-pedestrian), or V2X (vehicle-to-everything), is pending a rulemaking decision by NHTSA to establish performance standards. NSC encourages NHTSA to release this standard soon so that implementation of V2X can be more widespread.

DSRC can create redundant safety systems in motor vehicles. In other modes of transportation, fail-safe designs can support operator error, but in highway vehicles that task has fallen solely on drivers. DSRC would allow a vehicle to communicate with a red light to compensate for a fatigued driver, stop a car to prevent a collision with a pedestrian if a driver fails to detect him or her, and prevent or mitigate collisions between vehicles equipped with DSRC. DSRC has been deployed by some manufacturers, but NSC believes it is an important option in a safe systems approach to the design of HAVs and anticipates it will be more widely deployed if there is more regulatory certainty.

Education and Training

One component in the **National Highway Traffic Safety Administration's Automated Vehicle (AV)** policy, released last year, that should be a requirement moving forward is the incorporation of driver education and training about new safety technologies. With nearly 17.4 million new passenger cars and trucks sold in 2015,[12] understanding the technology on these vehicles is necessary, yet a University of Iowa survey found that 40 percent of respondents reported they had experienced a situation in which their vehicle acted in an unexpected way.[13] When this occurs in a real-life driving situation, among multiple drivers, it can lead to disastrous outcomes.

The National Safety Council and our research partners at the University of Iowa are focused on educating consumers about in-vehicle safety technology through our *MyCarDoesWhat*

---

[12] http://www.autoalliance.org/auto-marketplace/sales-data
[13] University of Iowa. National Consumer Survey of Driving Safety Technologies. July 30, 2015. Accessible at http://ppc.uiowa.edu/sites/default/files/national_consumer_survey_technical_report_final_8.7.15.pdf

campaign.[14] This brand agnostic education campaign informs drivers about how safety technologies work, how to best interact with them, and how to identify situations when the technology may not perform optimally and should not be relied upon. Because of the need for continued human involvement in the operation all vehicles today, the campaign tagline is *You are your car's best safety feature*.

Visitors to MyCarDoesWhat.org realize improvement in general knowledge and accurate comprehension of vehicle safety features. Drivers cannot effectively use these life-saving technologies if they do not understand both their functions and limitations. The AV policy proposes that this education be delivered in multiple ways, including computer based, hands-on and virtual reality training, and other innovative approaches. The *MyCarDoesWhat* education campaign follows that approach, and is developing virtual reality modules for release early next year. Further, we recommend ongoing evaluation to determine the effectiveness of the various messages, methods of delivery and media so they can be improved over time.

Standardized Nomenclature and Performance Outcomes

Another way to reduce consumer confusion is to standardize the nomenclature or taxonomy for advanced technologies. NSC, the State of California, and Consumer Reports have recommended that, at the very least, systems that are not completely automated or Level 5 should not be described as such. ADAS, with emphasis on *driver assist*, represents the vehicles being sold today and requires drivers to remain fully engaged in the driving task. That fact is often lost in marketing, media reports and consumer expectations. Labeling a motor vehicle as **"autonomous" today, or even using terms such as "autopilot"**, only confuses consumers and can contribute to losses of situational awareness around the driving task.

By establishing standard nomenclature and establishing clear performance outcomes, consumers will better understand what they should expect from these technologies. For example, vehicles marketed as having AEB will not necessarily come to a complete stop before a collision.[15] Some AEB systems only operate at higher speeds, and some are designed to slow rather than stop prior to a collision. These nuances may not be easily understood by consumers. IIHS reports that systems with a warning only, but no automatic corrective action, reduce frontal crash rates by about 25%, but vehicles with automatic braking reduce crashes by more than 40%. Vehicles with a warning and automatic braking reduce crash rates by about 50%. Establishing a standardized, results-based, understandable definition of AEB and other ADAS technologies would benefit consumers, manufacturers, and dealers, as well as organizations that evaluate vehicles for their safety benefits.

Finally, the New Car Assessment Program (NCAP) program has operated for nearly 40 years with a goal of testing vehicle safety systems and educating consumers about them. Practically, it has created a mechanism to allow consumers to evaluate vehicles on safety systems. NSC supports NCAP and believes it is an important program to improve the safety of the motor vehicle fleet. Standardized nomenclature and performance outcomes will ensure NCAP can more effectively compare vehicle safety systems between manufacturers, and even between a **manufacturer's own models**.

---

[14] www.mycardoeswhat.org
[15] http://www.nsc.org/learn/safety-knowledge/Pages/Driver-Assist-Technologies.aspx

American National Standards Institute (ANSI) Standard

As important as it is for the average consumer to know and understand the ADAS and automated technology, there is also work to be done on this issue as it relates to the technology and its rollout to commercial fleets. As such, NSC is taking a leading role working with the American Society of Safety Engineers (ASSE) and a wide array of experts in the automotive industry, technology sector, academia and fleet management, to develop an ANSI standard to address policies, procedures and management processes that will assist in the control of risks and exposures associated with the operation of autonomous fleet vehicles on public thoroughfares.

Road to Zero

On October 5, NSC, NHSTA, the Federal Highway Administration (FHWA), and the Federal Motor Carrier Safety Administration (FMCSA) announced the Road to Zero (RTZ) Coalition. RTZ is a partnership initiative focused on dramatic reductions in roadway fatalities. Over 80 public and private organizations attended the announcement to learn more about committing to a shared vision of zero fatalities on our roadways. The first meeting of the coalition will be on December 15.

The purpose of the Road to Zero Coalition is to 1) encourage and facilitate widespread implementation of countermeasures to reduce motor vehicle crash deaths in the near term; 2) develop a scenario-based vision for zero US traffic deaths in the future; and 3) provide a roadmap for policymakers and stakeholders to eliminate traffic deaths.

NSC is joined on the Steering Group for the Road to Zero Coalition by the following organizations: Advocates for Highway and Auto Safety, American Association of Motor Vehicle Administrators (AAMVA), American Association of State Highway and Transportation Officials (AASHTO), American Automobile Association (AAA), Commercial Vehicle Safety Alliance (CVSA), Global Automakers, Governors Highway Safety Association (GHSA), Institute of Transportation Engineers (ITE), Insurance Institute for Highway Safety (IIHS), Intelligent Car Coalition, International Association of Chiefs of Police (IACP), Mothers Against Drunk Driving (MADD), National Association of State Emergency Medical Services Officials (NASEMSO), National Association of City Transportation Officials (NACTO), National Association of County Engineers (NACE), and the Vision Zero Network.

On behalf of the Coalition, the NSC is administering a grant program to support national non-profit organizations committed to roadway safety programs that address the overlaps and gaps between roadway users, vehicles and infrastructure. The first round of grants were awarded earlier this month to seven winners. In addition, the Coalition will look at engaging others in near term solutions and countermeasures to reduce the death toll on our roadways. Finally, we will also provide critical input for the development of a future community scenario with zero traffic fatalities—an effort to look at the measures, programs and technologies will be necessary to reach zero highway fatalities in thirty years and work back from there. NHTSA, FHWA, FMCSA, and NSC are sponsoring the development of the scenario-based vision for zero traffic deaths in the U.S. in a 30-year timeframe, and the RAND Corporation has been retained to produce the scenario over the next 12-18 months. I look forward to briefing this Committee and others in Congress on the results of these activities and the efforts of the Coalition to reach zero deaths on our roadways.

Conclusion

Today, we have millions of drivers behind the wheel, spend millions of dollars on education and enforcement campaigns, and still recognize billions in economic loses as a result of crashes. In spite of safer vehicle designs and record-setting seat belt use rates across the nation, operating a motor vehicle remains one of the deadliest things we do on a daily basis.

It will be a long time before HAVs replace our current fleet. The transition will likely be messy as we deal with a complex and ever-changing Human-Machine interface. There will be an evolution of the existing technologies and perhaps a revolution when it comes to new and different technologies. We need to be prepared for unanticipated consequences and new failure modes.

Although we can imagine a future with automated vehicles, it will be a long and winding road to get to the destination of zero fatalities as a result of HAVs. We cannot afford to ignore the carnage on our highways that is a national epidemic today. The US trails other industrialized countries in addressing highway deaths. Efforts like Road to Zero will decrease fatalities today, tomorrow, and in the future if we embrace proven countermeasures and accelerate deployment of effective ADAS technologies.

NSC appreciates **this Committee's** leadership on vehicle technology and safe roadway transportation. If safety for the traveling public is the ultimate goal, advanced technology provides the most promising opportunity to achieve that outcome, and will go a long way toward reaching the goal of eliminating preventable deaths in our lifetime.

Aston Martin ○ Ferrari ○ Honda ○ Hyundai ○ Isuzu ○ Kia
Maserati ○ McLaren ○ Nissan ○ Subaru ○ Suzuki ○ Toyota

Statement for the Record of John Bozzella
President and CEO, Association of Global Automakers, before the
House Committee on Energy and Commerce
Subcommittee of Digital Commerce and Consumer Protection
Hearing on "Self-Driving Cars: Levels of Automation."
March 28, 2017

On behalf of the Association of Global Automakers ("Global Automakers"), I am pleased to provide the following statement for the record to the House Energy and Commerce Committee Subcommittee on Digital Commerce and Consumer Protection hearing on "Self-Driving Cars: Levels of Automation."

Global Automakers represents the U.S. operations of international automobile manufacturers and automotive suppliers. Our automaker members design, build, and sell cars and light trucks in the United States and abroad, and these companies have invested $56 billion in U.S.-based facilities, directly employ nearly 100,000 Americans, and sell 47 percent of all new vehicles purchased annually in the country. Combined, our members operate more than 300 production, design, R&D, sales, finance and other facilities across the United States.

By convening this hearing concerning the levels of motor vehicle automation, the Subcommittee recognizes the importance of accurately defining what automated vehicles are and what they can do. The term "automated vehicle" encompasses much more than the "self-driving" car that garners so much press these days. Rather, vehicle automation is evolving and will eventually entail a range of functionality depending on customer needs and on the business model of the developer. While the future may see the deployment of fully driverless cars, several automated and connected technologies are helping to improve safety on our roads *today*. Features such as adaptive cruise control and lane-keep assist provide support to the driver by enabling vehicles to assume a greater portion of the driving task. These technologies—which are foundational to the development of more highly automated systems—significantly enhance motor vehicle safety and the driving experience. As these systems become more advanced, a vehicle's capability to operate without the active control by the driver will increase. Additionally, increased penetration

GlobalAutomakers

of automated features will raise consumer awareness and familiarity with these advanced technologies and can help smooth consumers' transition to greater levels of automation.

It is important for policymakers to recognize that vehicle automation can cover a range of capabilities. SAE International has established a "Taxonomy" for automated vehicles in its Standard J3016, which defines different levels of automation, ranging from Level 0 (meaning no automation at all) to Level 5 (where a car can drive itself in all conditions with no supervision or input from a human driver). The Department of Transportation's Federal Automated Vehicle Policy uses this Taxonomy, which Global Automakers supports. The use of uniform definitions that recognize the various levels of automation is critical to consumer understanding of the technology and consistent treatment of automated vehicles on our roadways.

Unfortunately, state legislatures and regulators do not always follow or consider the SAE Taxonomy in their efforts to regulate automated vehicles. Over the past several years, we have seen a number of state proposals using varying definitions for automated vehicles, and in many cases these definitions do not account for increased levels of automation. Indeed, in some instances the definitions are so poorly constructed that they would unwittingly ban the sale and operation of technology in cars on the road today.

This is why federal leadership on automated vehicle policy is so important. Decisions made today will determine how fast and how far our systems evolve, and inconsistent policy approaches—particularly as they relate to vehicle characterization, performance, and design—could have significant long-lasting impacts. The federal government must develop a framework to encourage the development of highly automated vehicles, and work with state and local policymakers to provide guidance and establish clear policy roles and responsibilities. In our view, the traditional lanes of federal and state responsibility should apply with respect to automated vehicles: the federal government is responsible for safety standards that impact the design and performance of motor vehicles while states handle matters such as driver responsibility, insurance, and registration.

We therefore believe that Congress should preempt state laws and regulations that prescribe design and performance standards for automated vehicles. This concept is deeply enshrined in the structure of the current Motor Vehicle Safety Act, which recognizes the limited role states

GlobalAutomakers

play in regulating the safe design of motor vehicles. This structure should be no different for advanced motor vehicle safety technologies that rely as much on computer software as they do on hardware. State preemption essential to ensure that automakers are not subject to conflicting state regulations that will undoubtedly slow the pace of innovation, and limit the ability to manufacture vehicles that can operate in all 50 states.

It is also important for policymakers to recognize that there is an additional aspect of automation that needs to be considered in order to support the safety and mobility benefits of these emerging technologies: vehicle connectivity. Cars equipped solely with sensors such as radar, LIDAR and cameras will have limitations on the way in which they can sense the surrounding environment and interact with the vehicles on the road around them. Dedicated Short Range Communications (DSRC) supports "vehicle to everything" (V2X) communications, allowing cars to wirelessly connect to other road users and the surrounding infrastructure to effectively "see" around corners and through vehicles to achieve greater 360-degree situational awareness. DSRC can work alone as a sensor to inform or warn the driver to avoid a crash, or it can work in concert with other sensors and vehicle systems to support automated driving features, such as cooperative adaptive cruise control.

Soon, we will begin sharing our roads with automated vehicles. DSRC is the code that connects the automated vehicle world together—the communication standard that will ensure, regardless of mode or automation, vehicles are talking to each other, and that they are speaking the same language. This technology is already on the road today and ready for widespread deployment. The Department of Transportation has recognized the life-saving potential of DSRC and has issued, with the support of vehicle manufacturers, a proposed rule to create a new safety standard that would require new vehicles to be equipped with this technology and transmit a wireless basic safety message to support vehicle-to-vehicle (V2V) communications.

While the V2V rulemaking is a critical step in the right direction, it is important to understand that DSRC technology will create a new wireless transportation application ecosystem that will enable safer, smarter, and more efficient travel. DSRC operates over the 5.9 GHz spectrum band, which is important as it supports the low-latency communications needed for DSRC vehicles to speak to each other and the surrounding infrastructure (at the rate of ten messages per second).

GlobalAutomakers

The safety-critical applications in development throughout the DSRC band will support not only V2V, but also vehicle-to-infrastructure and vehicle-to-pedestrian communications, as well as DSRC applications to support automated features and highly automated driving. Finalizing this proposed rule and protecting DSRC from harmful interference in the 5.9 GHz band will provide the transportation industry, including original equipment manufacturers and aftermarket suppliers, with the necessary federal standards and certainty needed to increase deployments of this revolutionary technology.

Global Automakers and our member companies believe that connected and automated vehicle technologies can provide significant benefits. If policymakers can ensure an environment where innovation is permitted to thrive, connected and automated vehicles can truly transform the way we move goods and people. Already automated and wireless connectivity have revolutionized our economy, and bringing these innovations further into the transportation sector will be no less groundbreaking. Congress must work with all stakeholders to ensure that we have consistent rules for the deployment of automated and connected vehicles of all levels, and that the necessary policies are in place to support continued investment and education so that the labor market can adapt over time to support the new jobs and opportunities created by these new technologies.

Before the
DEPARTMENT OF TRANSPORTATION
National Highway Traffic Safety Administration

In the Matter of)
)
Federal Automated Vehicles Policy) Docket No. NHTSA-2016-0090
)

**COMMENTS OF THE U.S. CHAMBER OF COMMERCE
TECHNOLOGY ENGAGEMENT CENTER**

Highly Automated Vehicles ("HAVs") have tremendous potential to make our roads safer, enhance worker productivity, increase transportation efficiency, and deliver numerous other societal benefits. With its deep automobile manufacturing experience and capability along with world-leading information, communications, and technology ("ICT") companies, the United States has the potential to define and lead this new life-saving innovation and business opportunity. NHTSA's recently issued Notice and Request for Comments on the Federal Automated Vehicles Policy (the "Guidance") opens the dialogue on the appropriate national framework to facilitate the investment and innovation necessary for U.S. companies to claim a leadership role in this new era of transportation. The U.S. Chamber of Commerce's Technology Engagement Center ("C_TEC") welcomes the opportunity to provide input to NHTSA on this important topic. The Chamber established C_TEC to be the face of technology in the economy and to advocate for rational policy solutions that drive economic growth, spur innovation, and create jobs through the backing of a leading national and global business organization.

C_TEC appreciates that the Guidance is just that—guidance seeking voluntary cooperation—and not a proposed rule. NHTSA should maintain that status until more is known about this new transportation opportunity. As a first principle, C_TEC believes that government actions in this new area must be grounded in the realities of the complex and fast-changing HAV public testing and development process in order to foster rather than hinder investment and innovation. As C_TEC member Intel recently testified before Congress, "[our] vision for the future of transportation is one of zero accidents, mobility for all, environmental sustainability, reduced congestion, increased efficiency and *innovation that evolves at the pace of technology to ensure U.S. global leadership.*"[1]

Other agencies recently have faced similar experiences with fast-paced marketplace innovation, and NHTSA can learn from their efforts. Specifically, C_TEC urges NHTSA to reference the Federal Aviation Administration's ("FAA") new Small Unmanned Aircraft Rule as a guide.[2] The FAA also had the hard job of crafting regulations for a new form of transportation still in development. After the agency engaged with manufacturers and other interested parties, its final rule, which went into effect this summer, reflects a flexible approach that will allow industry to continue testing and developing new forms of transportation and experimenting with new operational environments while still promoting safety. In a similar vein, in considering its own role NHTSA should have the following goals for the Guidance: it should be supportive of innovation and market competition, informed by experts, technologically neutral, and promote coordination across the States.

---

[1] Prepared Statement for the Record of Intel Corporation, U.S. Senate Hearing on "How the Internet of Things Can Bring U.S. Transportation and Infrastructure into the 21st Century," at 4 (June 28, 2016), https://www.commerce.senate.gov/public/_cache/files/46c728ce-377e-4060-9cac-55db2230ddf8/17D163EB418271C1D3BBC8D572D589EE.doug-davis-testimony.pdf.

[2] *See* 14 C.F.R. § 107.

I. **The Guidance Raises Critical Questions for Which the Department Should Continue to Solicit Expert Feedback**

To aid NHTSA in ensuring HAV safety, the Guidance requests that manufacturers and other entities voluntarily submit a Safety Assessment Letter indicating how they have considered elements of the Guidance in testing and deploying HAVs. The Safety Assessment proposes to cover a wide variety of areas, ranging from technical topics such as data recording and cybersecurity to traditional regulatory areas such as registration and certification.[3] C_TEC commends NHTSA for the next steps it has laid out in the Guidance: holding public workshops and engaging experts for their review.[4] C_TEC urges NHTSA to work with the broad automated vehicle industry in an in-depth manner to clarify elements of the Guidance and determine if there are alternate solutions to technical issues that would more realistically align with the innovation, development, and testing processes for HAVs. We call particular attention to the following:

A. *Timing*

The Guidance expects that manufacturers and other entities will provide the Safety Assessment four months before active public road testing of any new automated feature.[5] Imposing such an expectation before testing will slow down American companies' development of this technology in a fast-paced global HAV market, where new technological developments are introduced for public road testing on a daily basis. Both safety and timeliness should be critical elements of the Guidance process.

---

[3] U.S. Department of Transportation and National Highway Traffic Safety Administration, *Federal Automated Vehicles Policy: Accelerating the Next Revolution in Roadway Safety* at 15 (Sept. 2016), https://www.transportation.gov/sites/dot.gov/files/docs/AV%20policy%20guidance%20PDF.pdf ("Federal Automated Vehicles Guidance").

[4] *Id.* at 34.

[5] *Id.* at 16.

B. Data Recording and Sharing

The Guidance proposes an anonymized data collection and sharing regime, but does not consider the difference between different types of data. NHTSA recommends that manufacturers and other entities collect data associated with fatalities, injuries, and vehicle damage as well as data associated with positive outcomes. Not only should this data be "readily available for retrieval by the entity itself and NHTSA," but each company should also "develop a plan for sharing its event reconstruction and other relevant data with other entities."[6] There are significant benefits to data sharing, but in order to promote a voluntary regime that companies will follow, they must be balanced against the costs of gathering data and sharing it with competitors. Is sharing positive outcome data as beneficial to the industry as a whole as sharing crash data? What types of data are proprietary? Would data be shared without repercussions? Industry must be further consulted and should lead in the development of any data-sharing plan to answer these questions as well as others.

C. Privacy and Cybersecurity

NHTSA does not provide prescriptive requirements for privacy and cybersecurity in the Guidance. This is the correct approach for several reasons. First, companies in this sector have long experience with promulgating privacy policies and best know their relationships with their customers. Changes to these policies should be informed by customer expectation and technological development. Therefore, companies should lead in developing new requirements, as they best understand both the technology they are creating and their customers who will use it. Second, with respect to cybersecurity, the Guidance correctly notes that this is an "evolving area

---

[6] *Id.* at 18.

and more research is necessary before proposing a regulatory standard."[7] C_TEC actively monitors developments in this area, and agrees that implementing in-house cybersecurity programs is vital for industry participants. We thank NHTSA for its recognition in the Guidance of industry efforts like the Auto Information and Sharing Analysis Center ("Auto-ISAC").

II. Pre-Market Approval Would Jeopardize Private Sector Innovation

In discussing potential new regulatory tools that NHTSA may use in this area, NHTSA indicates that it is considering pre-market approval authority. The Guidance acknowledges that pre-market approval is not currently part of NHTSA's authority, and qualifies that its discussions of how it could be implemented are preliminary.[8] However, in these preliminary discussions, NHTSA not only proposes an inflexible government pre-clearance process, but also suggests that as part of this it could implement a "technology-specific process to vehicles that include lower levels of automation, below L3-L5."[9] This proposal reflects a misguided policy and a radical departure from current practice, and should not progress pass the preliminary stage of discussion.

First, it is unwise for the Department to adopt a technology-specific process targeted at HAVs. Doing so would run the potential risk of the U.S. government indirectly picking technology winners and losers by approving one manufacturer's system quickly while delaying another's time to market. As C_TEC and its members have previously emphasized in other proceedings, "standards need to be voluntary and carefully designed so they do not constrain innovation."[10] Commenting on the role of government in fostering the advancement of the

---

[7] *Id.* at 21.

[8] *Id.* at 72-74.

[9] *Id.* at 72.

[10] Comments of the U.S. Chamber of Commerce Center for Advanced Technology and Innovation, Docket No. 160331306-6306-01 at 10 (June 2, 2016). The Chamber's Center for Advanced Technology and Innovation is C_TEC's predecessor.

Internet of Things, C_TEC member Qualcomm noted that "much is unknown about the future uses" of that technology.[11] The same is true of HAVs. "For this reason, the U.S. Government should tread very carefully in the legislative and regulatory space to let any and all innovative, and potentially ground-breaking, technologies be freely developed."[12] Technological neutrality will benefit consumers, who will receive the benefits of new technologies faster, and American manufacturing, which will be able to flex its muscles and develop devices, applications, and services that are currently unimagined.

Second, and in a similar vein, pre-market approval could harm competition. If NHTSA were not up to speed on the newest technologies—an expertise that would be difficult for the agency to grow and maintain, given the fast pace of the industry innovation—it could prevent new life-saving technologies from entering the marketplace and even risk negative impact on one company's competitiveness or in fact the whole U.S. industry's competitiveness. A responsive and flexible regulator gives manufacturers the ability to quickly perfect critical new tools, both protecting consumers and keeping the industry competitive.

Third, NHTSA has failed to justify a radical departure from its current self-certification and compliance testing process. That process, coupled with NHTSA's safety programs and general regulatory authority, has been in place for many years and, as the Guidance proudly proclaims, is "responsible for saving hundreds of thousands of lives."[13] The Guidance fails to explain why the new technologies surrounding HAVs justify such a radical change from this long-standing and successful process. As has long been the practice, the government should

---

[11] Comments of Qualcomm Inc., Docket No. 160331306-6306-01 at 16 (June 2, 2016).
[12] *Id.*

consider defining objective performance standards for automobiles, and leave to manufacturers the important job of meeting those obligations and certifying compliance.

III. Consistent State Laws and Clarity in Interpretation of Federal Laws Are Beneficial to Innovation

The Guidance speaks both to proposed state laws and enforcement of federal laws. While C_TEC commends NHTSA on its proposed model state policy, we raise several issues concerning statements in the Guidance regarding existing federal laws and regulations. NHTSA proposes a model state policy that, if adopted, would create a "consistent, unified national framework for regulation of motor vehicles with all levels of automated technology, including HAVs."[14] C_TEC strongly supports this goal and welcomes the formation of a NHTSA working group for information sharing with states engaged on HAV issues. Avoiding a patchwork of state laws enables manufacturers to be more innovative and ensures that any safety-enhancing automated vehicle technologies will be available throughout the country.

Additionally, NHTSA should have a consistent message that the voluntary guidance is not a set of objective tests or criteria to which HAVs should certify. To effectively support the development of HAVs and the safety benefits they will bring, it is important to avoid creating conflicts between NHTSA's role as regulator of vehicle safety performance and the states' interests in advancing the safe operation of vehicles on their roadways. Section II of the Guidance suggests a pre-approval approach for the use of HAVs within the states, which may have the potential to encourage states to regulate the safety equipment of HAVs. We disagree with state-level pre-approval of vehicle safety performance because it would create conflicts

---

[13] Federal Automated Vehicles Guidance at 5.
[14] *Id.* at 37.

between NHTSA's role and the states' role and cause state laws and regulations to be unnecessarily complex, time consuming, and costly.

The Guidance also addresses NHTSA's enforcement authority, both existing and proposed. NHTSA has streamlined and expedited its process for evaluating and responding to information requests—a simple interpretation request that appears to improve safety and follows the Guidance will receive a response within 60 days, while a more complex request will receive a response within 90 days.[15] Similarly, NHTSA has expedited its process for exemptions and has committed to grant or deny petitions within six months. It has also proposed that Congress lift exemption limits (vehicle volume and time) on innovative HAV safety technologies. C_TEC welcomes these expedited "shot clocks" for agency action and revisions to remove unnecessary regulatory burdens. They are especially important in this fast-moving and competitive sector, and as we move forward to a new Administration routine agency actions should not be delayed.

Somewhat troubling are the vague statements in the Guidance concerning NHTSA's enforcement authority. The Guidance states that NHTSA has "broad enforcement authority under existing statutes and regulations to address existing and emerging automotive technologies."[16] NHTSA made the same statement in its recent Enforcement Guidance Bulletin, which post-dates the Federal Automated Vehicle Guidance.[17] In that Bulletin, NHTSA stated that its "enforcement authority concerning safety-related defects in motor vehicles and motor

---

[15] *Id.* at 54.

[16] *Id.* at 50.

[17] *See* NHTSA Enforcement Guidance Bulletin 2016-02: Safety-Related Defects and Automated Safety Technologies (Sept. 20, 2016), https://www.transportation.gov/sites/dot.gov/files/docs/12507-AV_site_FedReg_Final-Defects-Authority-Enforcement-Bulletin_2016.0....pdf.

vehicle equipment extends and applies equally to current and emerging safety technologies."[18] NHTSA acknowledged that the increased use of "electronic systems (such as hardware, software, sensors, global position systems (GPS) and vehicle-to-vehicle (V2V) safety communications systems) . . . may raise new and different safety concerns," but stated that this does not limit NHTSA's authority.[19]

The proposals outlined in the Guidance should only focus on the upper levels of automation (*i.e.*, L3-L5). Lower levels of automation are currently in the marketplace, have a proven success in helping consumers avoid accidents, and are currently regulated by NHTSA through its existing investigation and defect authority. C_TEC respects the role NHTSA has played in motor vehicle safety for nearly 50 years. With respect to upper levels of automation, we urge NHTSA to consider that regulation in this area may need to proceed differently than in the past. For example, HAV manufacturers can now push software updates to their customers over the air. Would NHTSA hold these fixes back from consumers until it analyzed them with a simulator? Inserting a period of delay before a software fix can be deployed is contrary to public safety and eliminates the very safety benefit the ability to remotely receive a software update creates. NHTSA itself admits that "[g]iven the newness of HAVs and the private sector demand for persons with the necessary types of scientific expertise to work with those technologies, there is a shortage of suitable candidates to meet the Agency's critical hiring needs."[20] Moreover, because NHTSA may require additional expertise—particularly with respect to hardware and software development—C_TEC urges the Office of Personnel Management to work with

---

[18] *Id.*

[19] *Id.*

[20] Federal Automated Vehicles Guidance at 82.

NHTSA on providing the agency with the special hiring tools necessary to source adequate industry expertise.

IV. Conclusion

Automated vehicle technology and self-driving cars have the power to change our economy and society in positive ways that we cannot even conceive of now, potentially creating benefits both for consumers and American industry. But this technology will only be able to flourish to its fullest extent if industry is allowed to safely develop, experiment, and test without being slowed by burdensome regulation. We urge this and the next Administration to utilize a flexible and rational guidance approach that aligns with the realities of the innovation development and testing process, so that America may lead the world in this highly competitive, life-saving automotive evolution.

Respectfully submitted,

/s/
Tim Day
Senior Vice President, C_TEC
U.S. CHAMBER OF COMMERCE
1615 H Street, NW
Washington, DC 20062-2000

November 22, 2016

Statement of American Car Rental Association

to the

House Energy and Commerce Committee's

Subcommittee on Digital Commerce and Consumer Protection

Hearing on

"Self-Driving Cars: Levels of Automation"

March 28, 2017

The American Car Rental Association (ACRA) respectfully submits this statement for the record of the House Energy and Commerce Committee's Subcommittee on Digital Commerce and Consumer Protection's hearing on "Self-Driving Cars: Levels of Automation" on Tuesday, March 28, 2017.

ACRA is the national representative for over 98% of our nation's car rental industry. ACRA's membership is comprised of over 300 car rental companies, including all of the brands you would recognize such as Alamo, Avis, Budget, Dollar, Enterprise, Hertz, National and Thrifty. ACRA members also include many system licensees and franchisees mid-size, regional and independent car rental companies as well as smaller, "mom & pop" operators. ACRA members have over two million registered vehicles in service, with fleets ranging in size from one million cars to ten cars.

ACRA's members strongly support the development and gradual deployment of "Highly Automated Vehicles" (HAVs) to improve transportation safety and reduce property damage and personal injury and deaths associated with vehicle accidents. However, the introduction of HAVs is a complex technical and public policy challenge. This challenge will require policymakers to address and incorporate existing safety, consumer protection, privacy and liability issues into a changing vehicle populations that includes HAVs -- while at the same time maintaining flexibility to accommodate new and evolving legal issues unique to HAVs that may not be apparent today.

The members of the American Car Rental Association (ACRA) purchase one out of every nine new cars sold in the United States each year – almost 2 million vehicles in 2016. To the extent that HAVs are introduced into the private passenger motor vehicle fleet in the next decade, ACRA members will be at the forefront of HAV deployment and on the front lines of the education of car rental customers with respect to interacting with HAVs safely.

The Promise and Challenges of Autonomous Vehicles – The widespread introduction of HAVs promises to reduce the number of deaths (about 40,000/year in the United States) and injuries (hundreds of thousands every year in the United States) caused by motor vehicle accidents – over 90 percent of which are caused by human error. But this promise is not without challenges in many complex areas, including thorny public policy issues that have been debated by many interested parties for decades, including:

- **Liability** – Federal and State liability statutes generally hold the driver of a motor vehicle liable for injuries and property damage caused by that driver's negligence. With respect to HAVs, there is no "driver" per se and thus determining responsibility for injuries and other harm become problematic. Federal and State policymakers should consider assigning liability for accidents caused by HAVs to the entities most capable of addressing design and functionality shortcomings in HAVS – in most cases, the vehicle and software designers and manufacturers, rather than the humans occupying the vehicle or the fleet owners.

- **Ownership of Motor Vehicles** – As we move towards an era of widespread HAV deployment, our notions of motor vehicle ownership likely will undergo a revolutionary change. Instead of owning our personal automobile, or renting a minivan for a family vacation, or boarding a bus for a ride to school, or hailing a taxi – all of these activities may be undertaken with different types of HAVs which may or may not be owned by an individual, a school district, or a fleet operator. Resolving vehicle ownership issues, including maintenance, accident reporting, data recording and sharing, and other heretofore unaddressed issues with respect to HAVs will need to be discussed and resolved.

- **Taxes and Fees** – HAVs hold the promise of eliminating distinctions between rental cars, taxis, ride-hailing services and individual motor vehicle ownership. With the introduction of HAVs, responsibilities must be apportioned for paying Federal and State motor fuel excise taxes, State and Local fees on car rentals, ride-hailing services and taxis, and State and Local vehicle registration and sales taxes.

- **Harmonization** – The customers of ACRA members cross state lines in their current rental cars without restrictions and likely will anticipate the ability to do the same with respect to HAVs rented from ACRA members. As a result, a myriad of complex and perhaps contradictory State laws or regulations with respect to technical, safety or operational standards for HAVs should be avoided wherever possible. Continued State regulation of HAVs in traditional areas such as licensing, registration and insurance requirements would not in most instances pose impediments to the introduction of HAVs in ACRA's opinion.

- **Privacy** – Federal and State regulators have started to wrestle with the difficult challenges of maintaining individual privacy with respect to data generated by today's increasingly complex and technologically advanced motor vehicles and promoting transportation safety and enforcement of Federal and State laws. Such thorny privacy issues will only be multiplied with HAVs, and ACRA urges policymakers to preserve the right of vehicle owners to control and own the data generated by HAVs.

- **Cybersecurity** – The increased automation of motor vehicles, leading ultimately to deployment of HAVs, heightens the risk of cyber-attacks on single cars or groups of vehicles. Such risks must be managed by vehicle manufacturers and designers. However, the same technology that opens HAVs to cyber-attacks may hold the promise of reducing motor vehicle theft and other crimes involving vehicles. The cybersecurity issues related to HAVs must be balanced between protection of the vehicle's occupants and aiding law enforcement agencies in crime prevention and the apprehension of criminals.

Thank you for the opportunity to present this statement for the record at this hearing. ACRA stands ready to work with the members of the Subcommittee and all State and Federal legislators and regulators, as well as the many stakeholders interested in the development and introduction of HAVs, in the months and years ahead.

Please contact Greg Scott with questions regarding ACRA's HAV development and deployment positions at 202-297-5123 or gscott@merevir.com.

Statement for the Record
Dan Galves, Senior Vice President, Chief Communications Officer
Mobileye

U.S. House of Representatives
Committee on Energy and Commerce
Subcommittee on Digital Commerce and Consumer Protection
"Self-Driving Cars: Levels of Automation"
Tuesday, March 28, 2017

Dear Chairman Latta, Ranking Member Schakowsky, Members of the Subcommittee, thank you for holding a hearing on this important topic and for the opportunity to share my views with you.

Mobileye is a global leader in vision technology for advanced driver assistance systems (ADAS) and autonomous driving. Our core mission is to save lives and prevent injuries by reducing vehicle collisions.

Our ADAS technology (Level 1 and Level 2) is adopted by most of the world's major auto manufacturers and there are currently over 15 million vehicles on the road that utilize Mobileye technology. In addition to our technology being installed by the auto manufacturers at the factory level, Mobileye also offers aftermarket ADAS products which can be installed in any vehicle, making it an ideal solution for fleets looking to improve safety.

Beyond ADAS, which purpose is to avoid and/or mitigate vehicle collisions, Mobileye technology is being used by a variety of automakers as an important enabler for future Level 3-5 autonomous vehicles. These include a Level 3 program with Audi where production vehicles are expected to launch in second half of 2017, a Level 4 program with BMW to be launched in 2021, and a partnership with Delphi Automotive to create a Level 4 autonomous platform for use by many automobile manufacturers by 2019.

Mobileye's technology is centered around proprietary software algorithms deployed on a family of proprietary computer chips called EyeQ®. The technology turns raw camera data into usable information that a vehicle system can use to enhance safety and eventually, we believe, to drive autonomously. The technology performs detailed interpretations of the visual field in order to anticipate collisions with other vehicles, pedestrians, animals, debris, and other obstacles. Our products are also able to detect roadway markings such as lanes, road boundaries, barriers, and to read traffic signs and traffic lights. These products are validated by Mobileye and its customers to the highest level of accuracy and automotive-grade standards and have been proven over millions of miles of real-world driving conditions.

Mobileye's unique capability focuses on interpreting data from a camera fitted onto vehicles. We'd note that Mobileye has been able to demonstrate fulfillment of all ADAS functions with high accuracy through a single monocular camera sensor configuration, thereby reducing cost and simplifying tooling and packaging within the vehicle. For Level 3-5 vehicles, other sensors like radar and lidar will be necessary for redundancy; however, only a camera can identify both shape (i.e. vehicles, pedestrians, general objects, etc.) and texture (traffic sign text, lane markings, subtle road boundaries, traffic lights, etc.) which leads us to believe that camera will be the primary sensor in Level 3-5 vehicles.

The Problem

According to the U.S. Department of Transportation's National Highway Traffic Safety Administration, 35,092 people died and 2,443,000 were injured as a result of vehicle collisions on United States roads in 2015.[1] The number of deaths has increased by 7.2 percent from the previous year, the highest increase in over fifty years. Over 90 percent of vehicle collisions are attributed to human factors, including distracted driving, fatigue and drunk driving.

Moreover, according to a Boston Consulting Group (BCG) report titled *A Roadmap to Safer Driving Through Advanced Driver Assistance System*, "the cost [of vehicle collisions] to society totals about $910 billion annually, equivalent to roughly 6 percent of U.S. GDP." The BCG report further states that ADAS technologies alone have the potential to prevent 30 percent of all crashes, and together with fully autonomous vehicles, vehicle collisions could be reduced by 90 percent.[2]

We at Mobileye believe that the number of deaths and injuries caused by vehicle collisions is unacceptable and are working hard to significantly reduce and/or mitigate collisions through our investment in and development of innovative ADAS and autonomous vehicle solutions.

Different Levels of Automation

Mobileye's ADAS technologies--those installed by auto manufacturers at the factory as well as our aftermarket products--significantly reduce vehicle collisions, saving lives, preventing injuries, and reducing costs associated with collisions. There are also notable improvements in driver behavior which appear to result in substantial reductions in fuel consumption.

As Mobileye continues to innovate and deploy its ADAS technology, the company is leading the efforts in autonomous vehicle innovation and development.

There are five generally accepted levels of automation.

Level 0
- No automation. The human driver is in control. This includes the majority of vehicles on the road today.

---

[1] https://crashstats.nhtsa.dot.gov/Api/Public/ViewPublication/812318?_ga=1.154839546.33277525.1477595715
[2] https://www.mema.org/sites/default/files/MEMA%20BCG%20ADAS%20Report.pdf

Level 1 and Level 2: Advanced Driver Assistance Systems
- **Some automated features are introduced, like automatic braking, stability control and cruise control, but a human is still in charge.** Level 1 means the car can only work one automated system at a time, while Level 2 means that multiple automated functionalities can work in tandem. For example, Automatic Emergency Braking (AEB) plus Lane Keeping Assist (LKA).
- Mobileye technology supports many Level 2 systems on the road today. In 2018, introduction of EyeQ®4, the tri-focal lensed camera, and higher complexity algorithms will result in substantially higher ADAS functionality, as well as initial Level 3 systems.

Level 3 through Level 5: Autonomous Driving

Level 3 – Autonomous Under Certain Circumstances
- **Level 3 automation means the vehicle can take over all driving functions under certain circumstances.** The less-complex highway environment (all vehicles moving in the same direction, no pedestrians, no complex intersections) is the most logical circumstance.
- All major functions are automated, including braking, steering, and acceleration. **At this level, the driver can fully disengage until the vehicle tells you otherwise.** This is where the vehicle crosses into true "autonomous capability," and when technology begins to enable substantial benefits beyond safety, such as increased productivity.
- Going from Level 2 to Level 3 requires substantial increases in functional safety levels and system redundancies. In certain circumstances, the vehicle will need to ask the driver to re-engage. Since the driver cannot be assumed to take control instantaneously, the system will need to ensure safety for a period of time when the driver is still not engaged (for example, to pull over and stop if the driver does not respond to repeated requests to re-engage).
- Mobileye expects this additional redundancy to be covered by additional sensors like radar and lidar (for shape and object detection) and by Mobileye's REM™ localization map for identification of safe drivable paths and knowledge of intersections and other traffic signage or instructions.

Levels 4 & Level 5 – Fully Autonomous
- **Level 4 and Level 5 vehicles are autonomous in all situations and driving environments, not just "under certain circumstances," as in Level 3.**
- In Level 4 there does not have to be a driver because the vehicle is prepared for every situation and the human has moved from being the driver to just a passenger. **Level 5 vehicles will not have a steering wheel or other human-used vehicle controls.**
- We believe the initial deployment of this technology will be for "ride-sharing" fleets within confined areas.
- Driverless vehicles make the ride-share model much more cost-effective and compelling, as they eliminate the biggest cost of ride-share fleets: the driver. Initial deployment into ride-share fleets brings two other significant benefits: We expect initial usage will include a trained operator in the driver seat, to enable consumers to gain experience with the technology with the assurance of a trained operator monitoring the situation, and the

ability to generate real-world performance data in a safe way. Once enough data is generated, we would expect the regulatory framework to approve widespread usage.
- By using our crowdsourced REM™ Roadbook technology, the move from "somewhere" autonomy to "everywhere" autonomy is simply a switch of a button because the maps will be continuously updated everywhere, not only in confined, geo-fenced areas.
- Eventually, we will see auto companies scale-up from ride sharing to "shared ownership" where people and organizations share ownership of a car that can drive anywhere. This is even more transformative than ride sharing because it opens up completely new business models for transportation.
- In 2016 Mobileye announced its partnership with BMW and Intel intending to bring a fully autonomous vehicle into serial production by 2021.
- Also in 2016, Mobileye announced a partnership with Delphi Automotive to produce a turnkey autonomous driving system designed for rapid adoption by a variety of automakers.

Mobileye has ADAS programs with more than 25 automakers around the world, 5 programs for Level 3 Semi-Autonomous, and 5 programs for Level 4 Autonomous Vehicles, including with BMW and Delphi.

Path Forward

Autonomous vehicles and the benefits they will provide through reducing collision-related deaths, injuries, cost, and enhancing mobility to underserved communities such as the elderly and physically disabled individuals, are immense. Mobileye is fully committed to the vision of autonomous vehicles and are one of the leading companies working on making it a reality.

However, we have some years to go before fully autonomous vehicles are ubiquitous on U.S. roads and there is much more that can and should be done to save lives today without waiting for fully autonomous vehicles.

ADAS technologies already exist and have proven to significantly reduce collisions. Adoption rates are increasing but can and should be expedited with the help of prudent policy initiatives. Every day that goes by over 90 lives are lost in the U.S. due to vehicle collisions. Many lives can be saved by using ADAS technologies available today. It is simply unacceptable to not use proven technologies that are readily available today.

Moreover, ADAS technologies being used in vehicles today serve as fundamental building blocks for autonomous vehicles of tomorrow. They are paving the way towards autonomy in terms of technology validation and public perception and acceptance, all of which will be critical to the safe and widespread deployment of autonomous vehicles.

Policy Recommendations

ADAS and autonomous vehicle technologies could significantly decrease the number of vehicle collisions caused by human factors, reducing the amount of deaths and injuries. Legislators and regulators can and should play an important role in expediting the adoption of these technologies.

1) NHTSA should update and strengthen U.S. New Car Assessment Program (NCAP) standards to give ADAS technologies greater weight. To receive credit, ADAS technologies should be included in the standard fit of a new vehicle similar to seat belts, front airbags, and anti-lock brakes, not as optional equipment. The features must also verifiably meet the highest standards. Technologies that do not meet these standards should not receive credit under the system.

2) Federal Motor Carrier Safety Administration (FMCSA), should expedite the development and implementation of its "Beyond Compliance" program, as directed by the FAST Act, providing credits to motor carriers who take concrete safety measure, including the installation of advanced safety equipment.

3) The U.S. government should lead by example in promoting road safety by adopting ADAS technologies on government owned vehicles. For instance, USPS is about to replace its aging fleet with up to 180,000 new delivery vehicles. This and other vehicle procurements for government departments and agencies should require that ADAS technologies be included with the new vehicles. Vehicles that will be on the road for years to come and don't have ADAS technology should be equipped with aftermarket ADAS products.

4) Policymakers should implement a federal tax incentives program to encourage greater and faster adoption of ADAS technologies. At the very least, trucks carrying hazardous materials and other higher-damage risk vehicles should receive financial incentives to expedite adoption of ADAS technologies.

5) Auto manufacturers should be allowed to use ADAS features to meet Corporate Average Fuel Economy (CAFE) standards.

6) Policymakers should establish a single national standard for autonomous vehicle safety, preventing a patchwork of state laws. A patchwork of laws could prohibit autonomous vehicles from traveling from one state to another, stifling investment, innovation and deployment of these much-needed technologies.

Thank you again for holding this important hearing and for allowing Mobileye to submit this statement for the record.

epic.org

Electronic Privacy Information Center
1718 Connecticut Avenue NW, Suite 200
Washington, DC 20009, USA

+1 202 483 1140
+1 202 483 1248
@EPICPrivacy
https://epic.org

March 27, 2017

The Honorable Robert Latta, Chairman
The Honorable Janice Schakowsky, Ranking Member
U.S. House Committee on Energy and Commerce
Subcommittee on Digital Commerce & Consumer Protection
2125 Rayburn House Office Building
Washington, DC 20515

Dear Chairman Latta and Ranking Member Schakowsky:

For today's hearing on "Self-Driving Cars: Levels of Automation,"[1] we write to you again regarding the privacy and safety risks of self-driving vehicles. For more than a decade, EPIC has warned federal agencies and the Congress about the growing risks to privacy resulting from the increasing collection and use of personal data concerning the operation of motor vehicles.[2] In recent years, we have become increasingly aware of the threat to public safety of Internet-connected vehicles.[3]

This past weekend, Uber suspend the company's self-driving cars program in Arizona after one of their vehicles was in an accident with a traditional car in Arizona.[4] The Uber vehicle,

---

[1] *Self-Driving Cars: Levels of Automation* before the House Committee on Energy & Commerce, Subcommittee on Digital Commerce and Consumer Protection,
https://energycommerce.house.gov/hearings-and-votes/hearings/self-driving-cars-levels-automation.
[2] EPIC, Comments, Docket No. NHTSA-2002-13546 (Feb. 28, 2003), *available at* https://epic.org/privacy/drivers/edr_comments.pdf ("There need to be clear guidelines for how the data can be accessed and processed by third parties following the use limitation and openness or transparency principles."); EPIC, Comments on the Benefits, Challenges, and Potential Roles for the Government in Fostering the Advancement of the Internet of Things, Docket No. 160331306-6306-01 (June 2, 2016), *available at* https://epic.org/apa/comments/EPIC-NTIA-on-IOT.pdf; EPIC, Comments on Federal Motor Vehicle Safety Standards: "Vehicle-to-Vehicle (V2V) Communications," Docket No. NHTSA-2014-0022 (Oct. 20, 2014), *available at* https://epic.org/privacy/edrs/EPIC-NHTSA-V2V-Cmts.pdf; EPIC, Comments on the Privacy and Security Implications of the Internet of Things (June 1, 2013), *available at* https://epic.org/privacy/ftc/EPIC-FTC-IoT-Cmts.pdf; EPIC et al., Comments on the Federal Motor Safety Standards; Event Data Recorders, Docket No. NHTSA-2012-0177 (Feb. 11, 2013), *available at* https://epic.org/privacy/edrs/EPIC-Coal-NHTSA-EDR-Cmts.pdf; EPIC, Comments, Docket No. NHTSA-2004-18029 (Aug. 13, 2004); *available at* https://epic.org/privacy/drivers/edr_comm81304.html.
[3] Statement of Khaliah Barnes, hearing on the *Internet of Cars* before the House Committee on Oversight and Government Reform, Nov. 18, 2015, https://epic.org/privacy/edrs/EPIC-Connected-Cars-Testimony-Nov-18-2015.pdf; Statement of EPIC, hearing on *Self-Driving Cars: Road to Deployment* before the House Committee on Energy and Commerce, Subcommittee on Digital Commerce & Consumer Protection, Feb. 14, 2017, http://docs.house.gov/meetings/IF/IF17/20170214/105548/HHRG-115-IF17-20170214-SD012.pdf.
[4] Mike Isaac, *Uber Suspends Tests of Self-Driving Vehicles After Arizona Crash*, New York Times, Mar. 25, 2017, https://www.nytimes.com/2017/03/25/business/uber-suspends-tests-of-self-driving-vehicles-

EPIC Letter to House Energy & Commerce   1   Self-Driving Cars: Levels of Automation
Subcommittee on Digital Commerce & Consumer Protection   March 28, 2017

Defend Privacy. Support EPIC.

had a person in the driver's seat but was in self-driving mode, presumably "Level 3." The accident with the Uber vehicle highlights the risks of the self-driving mode as well as the dangers of having vehicles on the road with traditional vehicles.

This is not the first accident involving an autonomous vehicle. Late last year, a self-driving car failed to stop at a red light at a busy intersection.[5] A Tesla owner was recently involved in an accident when the autopilot failed recognize a lane shift in a construction zone, resulting in a collision with a construction barrier.[6]

These accidents should alarm the Subcommittee and the public, but they are only one of myriad issues with autonomous vehicles. Wide-scale malicious automobile hacking is no longer theoretical.[7] Although a full-scale remote car hijacking is certainly a serious risk to car owners and others,[8] hijacking is not the only risk posed by connected car vulnerabilities.[9] Connected cars leave consumers open to car theft, data theft, and other forms of attack as well. These attacks are not speculative; many customers have already suffered due to vulnerable car systems.

For example, criminals have exploited vulnerabilities in connected cars to perpetrate car "ransomware" scams, "where a car is disabled by malicious code until a ransom is paid."[10] According to one expert, computer criminals have installed malicious software in cars via USB drives used by mechanics for diagnostics and software updates. The software shuts down, or "bricks," the car unless and until the driver meets the criminal's demands. The expert even discovered a case where an entire fleet of vehicles was disabled by ransomware. She warns that criminals can also infect a car with malware remotely over the car's wireless connection.[11]

Car manufacturers should adopt data security measures. Early mitigation of threats to public safety may reduce auto fatalities, spur innovation, and result in safer vehicles.[12]

---

after-arizona-crash.html; Steven Overly, *Uber Self-Driving Car Flipped On Side In Arizona Crash*, Chicago Tribune, Mar. 25, 2017, http://www.chicagotribune.com/bluesky/technology/ct-uber-self-driving-car-crash-20170325-story.html.

[5] Mike Isaac & Daisuke Wakabyashi, *A Lawsuit Against Uber Highlights the Rush to Conquer Driverless Cars*, New York Times, Feb. 24, 2017, https://www.nytimes.com/2017/02/24/technology/anthony-levandowski-waymo-uber-google-lawsuit.html.

[6] Antti Kautonen, *Tesla Driver Blames Autopilot for Barrier Crash*, Autoblog, Mar. 3, 2017, http://www.autoblog.com/2017/03/03/tesla-autopilot-barrier-crash/.

[7] Brief of *Amicus Curiae* EPIC, *Cahen v. Toyota Motor Corporation*, No. 16-15496 (9th Cir. Aug. 5, 2016), *available at* https://epic.org/amicus/cahen/EPIC-Amicus-Cahen-Toyota.pdf.

[8] *See, e.g.*, Andy Greenberg, *Hackers Remotely Kill a Jeep On the Highway–With Me in It*, Wired (July 21, 2015), https://www.wired.com/2015/07/hackers-remotely-kill-jeep-highway/.

[9] *See* Bruce Schneier, *The Internet of Things Will Turn Large-Scale Hacks Into Real World Disasters*, Motherboard (July 25, 2016), http://motherboard.vice.com/read/the-internet-of-things-will-cause-the-first-ever-large-scale-internet-disaster (explaining that information systems face three threats: theft (i.e. loss of confidentiality), modification (i.e. loss of integrity), and lack of access (i.e. loss of availability)).

[10] Nora Young, *Your Car Can be Held for Ransom*, CBCradio (May 22, 2016), http://www.cbc.ca/radio/spark/321-life-saving-fonts-ransomware-cars-and-more-1.3584113/your-car-can-be-held-for-ransom-1.3584114.

[11] *Id.*

[12] *See generally*, Ralph Nader, *Unsafe at Any Speed* (1965).

EPIC urges this subcommittee to take these accidents and security flaws into account as you examine the various levels of automation in these vehicles. In addition to the substantial privacy concerns that these new cars present,[13] these recent incidents show that there are substantial safety concerns to everyone on the road.

Several states have recognized the risks to their residents and have passed laws regulating connected vehicles.[14] But consumer nationwide deserve protection. National minimum standards for safety and privacy are needed to ensure the safe deployment of connected vehicles.

We ask that this letter be entered in the hearing record. EPIC looks forward to working with the Subcommittee on these issues.

Sincerely,

Marc Rotenberg
Marc Rotenberg
EPIC President

Caitriona Fitzgerald
Caitriona Fitzgerald
EPIC Policy Director

Kim Miller
Kim Miller
EPIC Policy Fellow

---

[13] 8 U.S. Gov. Accountability Office, GAO-14-649T, Consumers' Location Data: Companies Take Steps to Protect Privacy, but Practices Are Inconsistent, and Risks May Not be Clear to Consumers (2014), http://gao.gov/products/GAO-14-649T; Jeff John Roberts, *Watch Out That Your Rental Car Doesn't Steal Your Phone Data,* Fortune, Sep. 1, 2016, http://fortune.com/2016/09/01/rental-cars-data-theft/.

[14] Nat'l Highway Traffic Safety Admin., Federal Automated Vehicles Policy (Sep. 2016), https://www.transportation.gov/sites/dot.gov/files/docs/AV%20policy%20guidance%20PDF.pdf ; Ark. Code § 23-112-107; Cal. Veh. Code § 9951; Colo. Rev. Stat. § 12-6-401, -402, -403; Conn. Gen. Stat. § 14-164aa; Del. Code § 3918; Me. Rev. Stat. Ann. tit. 29-A §§ 1971, 1972, 1973; Mont. Code § 61-12-1001 et seq.; Nev. Rev. Stat. § 484D.485; N.H. Rev. Stat. § 357-G:1; N.J. Stat. § 39:10B-7 et seq.; N.Y. Veh. & Traf. Code § 416-b; N.D. Cent. Code § 51-07-28; Or. Rev. Stat. § 105.925 et seq.; Tex. Transp. Code § 547.615; Utah Code § 41-1a-1501 et seq.; Va. Code §§ 38.2-2212(C)(s), 38.2-2213.1, 46.2-1088.6, 46.2-1532.2; Wash. Rev. Code §46.35.010. 62 Va. Code Ann. § 38.2-2213.1 (West).

Honda North America, Inc.
1001 G. Street, N.W. Suite 950
Washington, DC 20001
Phone (202) 661-4400

March 28, 2017

The Honorable Robert Latta, Chairman
Subcommittee on Digital Commerce and Consumer Protection
2125 Rayburn House Office Building
Washington, DC 20515

The Honorable Janice Schakowsky, Ranking Member
Subcommittee on Digital Commerce and Consumer Protection
2322A Rayburn House Office Building
Washington, DC 20515

Dear Chairman Latta and Ranking Member Schakowsky,

Thank you for this opportunity to share Honda North America, Inc.'s (Honda) views on the "Self-Driving Cars: Levels of Automation" hearing. Honda envisions a future society that enjoys a zero-collision mobility experience for all road users. Incremental steps towards automation are important in reaching this goal to ensure the technology is safely deployed and widely accepted and utilized by consumers.

Honda has been investing and manufacturing in the U.S. for more than 40 years. This includes $27 billion in parts and materials from 610 U.S. suppliers and Honda has invested $3.4 billion in its U.S. factories over the past four years alone. Our 12 manufacturing plants produce passenger vehicles, power equipment, and power sports products. The U.S. also hosts the global headquarters for HondaJet. Honda directly employs 30,000 Americans and has never laid off a permanent associate.

The auto industry is on a path toward widespread use of highly automated (autonomous) vehicles, and the SAE levels of automation outline steps along that path. Enabling progression along that path is the advent and increased use of connected vehicles. The enhanced situational awareness resulting from connections between vehicles, vehicles and pedestrians, vehicles and infrastructure, and vehicles and other road users, also known as V2X technology, is vital to reach the full safety benefit of automation.

In order to immediately reap the benefits of V2X technology, it is vital to ensure that Dedicated Short Range Communications (DSRC) technology is used as a shared platform by all road users. DSRC has the capability to expand a car's situational awareness, reducing or eliminating automobile crashes, and ultimately, saving lives. This is achieved by both expanding the operational range of vehicle based sensors and by providing important redundancies in algorithm based decision-making.

DSRC, developed to operate in the 5.9GHz Safety Spectrum, will provide a consistent, reliable platform for these life-saving technologies. We commend NHTSA for putting forth an NPRM to ensure that all vehicles include this common technology platform. Moreover, most of the auto industry stands ready to invest in DSRC once the rule is finalized and there is regulatory certainty for all vehicles.

Honda has several technologies under development for deployment using the 5.9GHz Safety Spectrum today, however deployment is delayed due to the lack of confidence in the priority status for DSRC technology on that spectrum which is currently under deliberation at the Federal Communications Commission. We must be able to provide these technologies to our consumers with complete confidence that they will work. To do so, DSRC technologies must be prioritized on the 5.9GHz Safety Spectrum.

Honda continues to investigate expanded uses of this technology to improve safety and mobility issues. We are conducting studies of Vehicle to Infrastructure, Vehicle to Pedestrian, Vehicle to Motorcycle as well as to other vulnerable road users. Furthermore, Honda is currently looking at technology that would allow cars equipped with DSRC technology to monitor the road and road users, and allow the car and driver to react even if those other road users do not use DSRC technology. This type of technology is a critical enabler and should be viewed as a complementary technology on the path toward a highly automated vehicle future.

While examining the five levels of automation, it is our hope that the Subcommittee will also look at the most expedient ways to allow innovation to flourish in this area, including utilizing DSRC technology in the 5.9GHz Safety Spectrum. Honda looks forward to continuing to work with the Subcommittee on this important issue.